*Cambridge Studies in Biological Anthropology*  4

# Human adult odontometrics

*Cambridge Studies in Biological Anthropology*

*Series Editors*

G. W. Lasker
Department of Anatomy, Wayne State University, Detroit, Michigan, USA

C. G. N. Mascie-Taylor
Department of Biological Anthropology, University of Cambridge

D. F. Roberts
Department of Human Genetics, University of Newcastle upon Tyne

*Also in the series*

G. W. Lasker *Surnames and Genetic Structure*

C. G. N. Mascie-Taylor and G. W. Lasker (editors) *Biological Aspects of Human Migration*

Barry Bogin *Patterns of Human Growth*

J. E. Lindsay Carter and Barbara Honeyman Heath *Somatotyping – development and applications*

# Human adult odontometrics

## The study of variation in adult tooth size

JULIUS A. KIESER

*Witwatersrand University, Johannesburg*

The right of the
University of Cambridge
to print and sell
all manner of books
was granted by
Henry VIII in 1534.
The University has printed
and published continuously
since 1584.

CAMBRIDGE UNIVERSITY PRESS

*Cambridge*
*New York   Port Chester*
*Melbourne   Sydney*

CAMBRIDGE UNIVERSITY PRESS
Cambridge, New York, Melbourne, Madrid, Cape Town, Singapore, São Paulo

Cambridge University Press
The Edinburgh Building, Cambridge CB2 8RU, UK

Published in the United States of America by Cambridge University Press, New York

www.cambridge.org
Information on this title: www.cambridge.org/9780521353908

First published 1990
This digitally printed version 2008

*A catalogue record for this publication is available from the British Library*

*Library of Congress Cataloguing in Publication data*
Kieser, Julius A.
Human adult odontometrics.
(Cambridge studies in biological anthropology; 4)
Bibliography: p.
Includes index.
1. Dental anthropology.   I. Title.   II. Series.
[DNM: 1. Odontometry. 2. Tooth – anatomy & history.
WU 101 K47h]
GN209.K54   1990   573′.6314   89-9939

ISBN 978-0-521-35390-8 hardback
ISBN 978-0-521-06459-0 paperback

*This book is dedicated to my beloved wife and friend Glynny, and to our four wonderful children, Annie, JJ, Daniel and David.*

# Contents

# Foreword

The story of the search for precision in the delineation of human structure goes back to Blumenbach (1775), Linnaeus (1758), Buffon (1749) and their precursors; while Magnus Hundt used the word 'Anthropology' in 1501. There are those who would look to the inimitable Aristotle as our founding father – who applied the name 'anthropologists' to students of Man. Some of us, who may be biased by our links with medical science, would even stake a claim for Hippocrates (who had lived all but seven years of his life when Aristotle was born), for did not the 'Father of Medicine' leave us *De Natura Hominis*, a kind of primer of ecological anthropology?

The measure of man has thus a long and honourable history. Odontometry means the measurement of teeth and our records of its history are less well documented. Here, truly, is an archive waiting to be chronicled by some young and eager disciple who senses, however dimly, the fructifying force of the historical approach.

Yet even if its roots be less readily discernible, the modern flowering of odontometrics is one of the marvels of recent biological anthropology. It is not so many years ago, and certainly well within my lifetime, that people who measured hominid teeth, past and present, were considered a trifle eccentric. They were looked at askance by their head-shaking colleagues who were inclined to ask: where are they heading? why are they collecting thousands of metrics with their carefully-sharpened vernier calipers and their tunnel vision? Truth to tell, some of the odontometric practitioners wondered themselves where they were going and cogitated earnestly over what wise propositions to put forth, *after mensuration*, as the hypotheses they were setting out to test!

This book shows what has been achieved from the simple (but not always so simple) measurement of just two diameters of the crown of each tooth. Julius Kieser has woven a wondrously intricate and infinitely ramifying pattern of analysis. He has taken us by the hand and led us through the complexities of modern biological concepts. We are shown how the skilled statistical manipulation of odontometric data bids fair to throw light on some of the most vexatious challenges of modern general and human biology.

This very broad canvas embraces tooth size and heritability, secular and evolutionary trends, sexual dimorphism, variability within and between tooth groups and species, field theory, epigenetic canalisation, compensatory interactions in the size of adjacent teeth, a marvellously stimulating analysis of the phenomenon boasting the title of fluctuating odontometric asymmetry, allometric relationships and much more besides. The book is informed by comprehensive and critical reviews of the literature, including reference to the author's own prolific and important original contributions, many of them written jointly – in a most productive symbiosis – with his statistician-friend, Dr Hennie Groeneveld.

Julius Kieser is a product of the University of the Witwatersrand Dental School. The grandson of a sometime Professor of Anatomy, he was early drawn to the Wits Anatomy Department: branching away from his dental studies temporarily, he completed a Medical BSc degree majoring in Anatomy. It was a natural and almost irresistible step for Dr Kieser to return to the Anatomy Department to undertake his researches towards the PhD degree and much of the contents of this book reflect studies he made in his doctoral project.

The breadth of Dr Kieser's reading, and his mastery of a staggering array of anthropological, evolutionary, embryological, orthodontic and statistical concepts shine through every page of this work. I have no doubt that students and specialists in all of these fields will find in these pages something to illumine, catalyse and stimulate, their thinking.

PHILLIP V. TOBIAS

*DSc (Witwatersrand), ScDhc (Cantab)*
*Head, Department of Anatomy and Human*
  *Biology,*
*University of the Witwatersrand Medical*
  *School, Johannesburg.*

# Preface

The aim in writing this book has been to provide an introduction to variation in human adult tooth size, with the emphasis on general principles rather than specific applications, and one moreover, that can be understood by students and researchers of no more than ordinary mathematical ability. While aiming to interest as wide a range of readers as possible, it is principally directed at practitioners and students of physical anthropology, oral biology, orthodontics and forensic science. In consequence, some will find less detail than they require and others more. Those who are specialists in their respective fields will encounter all they require of the general principles involved, but will find little guidance in the techniques of orthodontics, forensics, genetics or physical anthropology. Those for whom odontometrics forms part of a course in any of these subjects will find a good deal more detail than they require.

There have been remarkable advances in the field of human odontometrics in the past fifty years. I have had the benefit of these works, not only as sources of data, but also as sources of inspiration and stimulation. Particularly important in this respect have been the works of Dahlberg, Hanihara, Lundström, Moorrees and, most importantly, Garn. These men were exceptionally well qualified to compile and interpret in a meaningful way a vast body of information on our teeth that now exists. Having paid tribute to these five great scientists, I must do the same to other, earlier pioneers of at least equal stature, namely Hellman, Hrdlička, Mijsberg, Nelson, Selmer-Olsen and Shaw. Their outstanding contributions still remain basic sources of information to students, clinicians and research workers alike.

This book began life as a series of lectures given at the Witwatersrand University. Subsequent versions evolved through lectures given at the Universities of Alberta, Cincinnatti and Tennessee, and at the Field Museum, Chicago.

I am grateful for the constructive criticism of those who attended these lectures. I am specifically thankful to Geoff Sperber, Jeff Osborn, Tony Perzigian and Ed Harris for their hospitality and encouragement. At an early stage of my research I received helpful suggestions from Noel Cameron, Hertha de Villiers, Bill Evans, Alex Frazer, Philip Gingerich,

Stephen Gould, Carl Huether, Richard Liversidge, Brian Preston, Jeff McKee, Shelley Saunders, Holly Smith and Frikkie van Reenen. People who deserve special thanks for their detailed responses to earlier drafts of various sections of this book are Robert Corruccini, Stanley Garn, William Hylander, Yuji Mizoguchi, Grant Townsend and Bernard Wood.

My greatest intellectual debt is to Phillip Tobias who stimulated and nurtured my interest in human odontometry, and under whose supervision much of this work was completed. Special thanks are due to Hennie Groeneveld who introduced me to biostatistics, and to the Editors and anonymous referees of many Journals who offered numerous helpful suggestions and criticisms of the papers on which this book rests heavily. The manuscript was read thoroughly by Derek Roberts and Gabriel Lasker, whose comments and suggestions have been comprehensive and helpful. The staff of Cambridge University Press have been exceptionally supportive, helpful and patient. I thank particularly the publishing team of Alan Crowden, Robin Smith and Ruth Henderson.

In conclusion, I wish to thank my lovely wife Glynny without whose help, encouragement and prodigious editing, this text would not have been completed. Only those who know her will appreciate how much she has contributed to this book, and without her, how little I would have achieved.

# 1   *Introduction*

Teeth provide a wealth of information relevant to a great variety of biological problems; they therefore attract the attention of the zoologist, the human biologist, and many others. The fact that dental dimensions have fascinated such a wide range of scientists is reflected in the spectrum of journals that contains results of their endeavours. Hence, the major purpose of this book will be to unite these diverse data into one single, easily accessible source.

Tooth length and width constitute the most widely documented of anthropometric features. These two measurements provide significant information on such human biological problems as the genetic relationships between populations and human environmental adaptation. As an experimental system, they serve as a test-bed for craniofacial dental procedures, whilst as an evolutionary entity, they aid in solving questions of comparative anatomy and phylogeny.

Because of their durability, teeth and bones are the last structures to disintegrate after death. Hence they contribute an inordinate body of evidence for the evolutionary anthropologist, the archaeologist or the forensic expert to rely on. Dental material is much less destructible than osseous material, so that collections of proto- and prehistoric human remains have come to contain disproportionate numbers of teeth. For example, from his census of the skeletal material derived from fourteen African hominid sites, Tobias (1972) concluded that teeth constituted about three-quarters of the total sample. In fact, our knowledge of early humans has largely been built around our knowledge of fossil teeth. Forensic experts have come to rely heavily on dental evidence for the identification of the victims of natural and unnatural disasters. Dental data provide considerable evidence on evolution, on the relationships between human populations and between individuals, on variations in normal growth and the factors affecting them, on control of development, on genetic mechanisms and on models of inheritance.

Dental development in man commences at about four weeks' gestational age with the formation of the buds from which the deciduous incisors grow. Not only is the final shape and size of the tooth crown determined well before its eruption into the mouth, but once it is formed the only change

1

that is possible is the removal of tooth substance through wear, pathology or trauma. Because crown size is as it were frozen in time, the oral biologist may, by studying the final crown size and shape, attempt to elucidate retrospectively those developmental perturbations that may have affected odontogenesis. Although teeth and the jaws that house them share an early period of size increase, the precocious cessation of dental growth has come to be recognised as a potential source of disharmony between tooth size and jaw size. It is this disharmony that has provided the clinical and theoretical challenges on which orthodontists have thrived.

Teeth are the only hard tissues in the body that may be observed directly without radiographic or other noninvasive intervention. Modern techniques of making casts have provided physical anthropologists with accurate, permanent and easily obtainable records of the dentitions of large numbers of human populations. Metrical data derived from such collections, analysed by multivariate statistical techniques, have contributed greatly to our understanding of human variation. The abundance of dental metrical data has also benefited the geneticists who have used this information in their population and pedigree analyses. Finally, the analysis of dental size asymmetry has recently emerged as a useful tool for studies in public health and epidemiology. For all these reasons a reference work on human odontometry would be of benefit to a broad spectrum of professional and academic disciplines.

A book of this kind does not cover every aspect of human odontometrics. It relates to the history of the subject matter. There are, inevitably, omissions and certain topics receive more emphasis than others. This partly reflects the author's own interests and expertise, but is also attributable to the Editors' concern that the book should be concept-orientated rather than concentrating on specific populations or methods.

No excuse is needed for venturing across many disciplines and fields of interest in this book for the real biological world extends across disciplines and fields of interest. Whilst respect is due to scholars who have patiently expanded our knowledge of the tooth size of a single tribe or a small sample of diseased human beings, these efforts are useful essentially for the evidence they give on issues of general and comparative scope. For instance, our inability to say anything coherent about the phylogenetic causes of odontometric change does not arise from a shortage of studies on the subject. Rather, it results from a younger generation who have been taught never to threaten reality with theory, and from the difficulty of applying insight from one discipline to interpret data from another. More excavations, more specimens and, above all, more money are not necessarily the answer. Without a strategy aimed at bridging the gap between all the specialities concerned, additional trips and larger museum collections will not lead to a better understanding of human odontometrics. If nothing else, it is hoped that this book will go some way towards an amalgamation of the different fields of interest concerned with the study of tooth size in man.

A final comment on terminology: because humans, in common with some other extant primates, have lost their first two premolars, the premolars should strictly be referred to as P3 and P4. However, in the present work, the more usual P1 and P2 will be adhered to. The abbreviated notations used in this work will be as follows: U for maxillary and L for mandibular, I for incisors, C for canines, P for premolars and M for molars. Teeth are counted from front to back (mesiodistally). Hence, for example, the upper second molar becomes UM2.

# 2 Measurement of tooth size

## 2.1 Measurements of tooth diameters

The choice of a method of tooth size determination may have an important bearing on the results of an odontometric analysis. For instance, some authors advocated the use of maximum tooth measurements (e.g. Nelson, 1938; Moorrees & Reed, 1954; Tobias, 1967) whilst others preferred using the smaller, minimum, dimensions (e.g. Lavelle, 1972). Each of these measures may be appropriate for resolving a specific problem, yet none can be regarded as apposite for all possible analyses. Unfortunately methodological and terminological diversity have tended to obscure rather than clarify the field of dental mensuration. The buccolingual measurement (BL), for example, has been referred to as a thickness (Miyabara, 1916), a diameter (Nelson, 1938), a breadth (Selmer-Olsen, 1949; Schamschula, Barnes & Adkins, 1972) and a length (van Reenen, 1966). The mesiodistal measurement (MD), on the other hand, has been called a width (Miyabara, 1916; Seipel, 1946), a breadth (Selmer-Olsen, 1949; Oliver, 1960) and a length (Hrdlička, 1952; Hunter & Priest, 1960; van Reenen, 1966; Schamschula et al., 1972).

In an attempt to standardise the terminology, Wolpoff (1969) argued that, because almost all teeth have a greater anterioposterior than mediolateral component, this measurement is justifiably described as a length. A breadth then, is the measurement perpendicular to it. However, the meaning of Wolpoff's terminology is unclear, since there are an infinite number of directions in a saggital plane.

Unfortunately, a profusion of measurement techniques cloud the field of odontometry. Some of the earlier methods of dental mensuration are described below, followed by a comparison of those methods that are currently in use.

For the MD diameter, Muhlreiter (1892) advocated the use of a distance from contact point to contact point, measured on the buccal surface of the tooth. This method was more recently employed by Hunter & Priest (1960) who noted that since the widest portion of a tooth is usually toward the buccal surface, it follows that the measurement from contact point to contact point is not necessarily the greatest mesiodistal diameter. Twenty-four years after Muhlreiter, Miyabara (1916) defined the MD diameter of a tooth as follows: 'If we assume two planes tangential to the approximal

4

surface, width is defined by the distance between them.' p. 740. The BL diameter, according to Miyabara, is the tooth's thickness, measured perpendicular to the tangent to the dental arch.

One of the first investigators to raise the problem of measuring different tooth classes was Nelson (1938), who noted that 'the mesiodistal diameter of the crowns of incisors and cuspids was taken as the maximum dimension across the cutting edge. The mesiodistal diameter of the crowns of premolars and molars was, however, measured parallel to the median mesiodistal axis of the crown from contact point to contact point. The labiolingual diameter was obtained at right angles to the mesiodistal diameter and at right angles to the vertical median crown axis' (p. 263). Selmer-Olsen (1949), varying the recommendation of Muhlreiter (1892), suggested taking a mesiodistal dimension between the points of contact in normal occlusion, but not necessarily on the buccal crown surface. This technique has more recently been employed by Perzigian (1976).

Possibly the most widely used method of tooth size determination is that of Moorrees & Reed (1954). These authors suggested that the greatest MD dimension of the crown should be measured parallel to the occlusal and labial surfaces. The BL distance was taken as the greatest distance between the labial and lingual surfaces of the tooth in a plane perpendicular to that in which the MD dimension was measured. Investigators that have used the above method include Moorrees & Reed (1954), van Reenen (1966), Potter *et al.* (1981), Lukacs (1981, 1985), Kieser (1985) and Axellson & Kirveskari (1983). Another widely used method is that of Goose (1963), which is similar, but specifies that MD measurements should be taken between the contact points in normocclusion or where they would have been, in the case of malpositioned teeth. The BL dimension according to Goose (1963) is to be taken at right angles to the MD, without regard for the position of the crown. This method has since found acceptance in the works of Moss & Chase (1966), Goose (1967), Moss, Chase & Howes, (1967), Barnes (1969), Hinton, Smith & Smith (1980) and Jacobson (1982).

One of the most lucid descriptions of the MD and BL diameters is Tobias's (1967) elaboration of the method originally described by Koren-hof (1960). Tobias described the MD dimension as the distance between two parallel lines perpendicular to the mesiodistal axial plane of the tooth. This measurement, he argued, should be taken tangential to the most mesial and most distal points of the crown, along a line parallel to the occlusal plane. In effect, this means that the termini need not necessarily correspond to the points of interstitial contact. Tobias described the BL crown diameter as the greatest distance between the buccal and lingual surfaces of the tooth crown, with the arms of the caliper held parallel to the mesiodistal axial plane of the tooth and tangential to the buccal and lingual surfaces. Tobias further pointed out that the contact points of the two arms of the instrument do not necessarily have to lie in the same buccolingual plane: 'By placing each arm tangentially along the respective surface, I have measured the perpendicular distance between two planes parallel to that in which MD was measured, but the actual points of contact of the left and right arms with the tooth surfaces are not necessarily directly opposite each other . . .' p. 145.

In a clear departure from the foregoing methods, Schamschula and his co-workers (1972), suggested measuring the BL diameter first, with the caliper jaws parallel to the estimated long axis of the tooth, and the MD diameter (length) at right angles to the plane of the BL measurement. In a more recent departure from the classic descriptions of MD and BL as maximum distances, Lavelle (1972) has suggested the use of a minimum distance between mesial and distal contact points, measured parallel to the occlusal plane.

## 2.2    Other measurements

In addition to crown length (MD) and crown width (BL), the crown height may be measured in those teeth that do not show excessive occlusal wear. The most useful definition of crown height is that of Moorrees (1957), the distance between the tip of the mesiobuccal cusp and the deepest point of the cemento–enamel junction on the vestibular side measured along a line parallel to the long axis of the tooth.

Composite measurements are also useful, particularly in comparing the sizes of different teeth in the same arcade, as well as making comparisons between different geographic groups. One of the simplest and most useful of the compound measurements is the crown area, defined as the product of the MD and BL diameters. Although this measure provides an indication of the overall surface area of the tooth crown, it fails to account for differences in shape. Moorrees (1957) pointed out that the crown area is in fact inaccurate in that a cross-section of the crown parallel to the occlusal surface is not rectangular. Others have suggested that the crown area may be considered to be an accurate representation of the rectangle into which the tooth would have fitted (had it been elastic), and hence may be an effective method of calculating tooth size for comparative purposes (Jacobson, 1982).

Another composite measure is the crown module, defined as half the sum of the MD and BL dimensions (Shaw, 1931; Robinson, 1956). As a rough measure of crown form, the crown shape index may be determined as follows:

$$\frac{100\,BL}{MD} \text{ (Tobias, 1967).}$$

Loring Brace more recently introduced a simple measure of summary tooth size, which he considered to be of value in the comparison of large numbers of teeth in different populations (Brace, 1980). The summary tooth size, TS, is the mean area per tooth defined as:

$$TS = \Sigma(MD \times BL)/Nj,$$

where $j = I1$, $U12 \ldots UM3$; $LI1$, $LI2 \ldots LM3$, and $Nj =$ total number of measured teeth in each category.

## 2.3 Reliability

A dental measurement, be it linear or composite, is only as valuable as its reliability. Reliability is the extent to which a measurement and its attendant technique are consistent. There are two very important properties of reliability: precision and accuracy. Precision refers to the repeatability of a measurement. The more precise the method, the more closely clustered about an average value will be the set of measurements. Accuracy, on the other hand, is an estimate of how close the measured value is to the true value, that is how unbiased is the measurement. A measurement may be very precise and yet very inaccurate. One could measure the width of a tooth with great precision, to the nearest tenth of a millimetre, yet if the calipers were faulty, one would have a precise, but inaccurate measurement.

For example, suppose we have a standard, manufactured, metal tooth of a known MD diameter of say, 12.4 mm. We then ask three observers to measure its MD diameter twenty times. Now it is a known fact that variation is an inherent property of every population. Yet a distribution of values may also be obtained by repeatedly determining the same measurement in a single object. This variation in the observed values may be caused by systematic or random errors which occur during the measuring process. Hence, in our example (Figure 2.1), the histograms generated for each of our three observers, demonstrate a scatter of results.

The relationship between the mean of a set of measurements and the true measurement gives an idea of the accuracy of the measurement process. Observer (a) showed a definite tendency to underestimate the true MD diameter of the standard metal tooth. Hence his error is systematic (e.g. his caliper may have been incorrectly zeroed), and his estimate is biased. However, observers (b) and (c) were accurate (unbiased) observers because their determined means are similar to the true mean.

Precision, on the other hand, is an estimate of the repeatability of our measurement process and may be described by the standard deviation of the scatter of measurements. The better the measurement technique, the fewer random errors will be recorded and the more closely will the

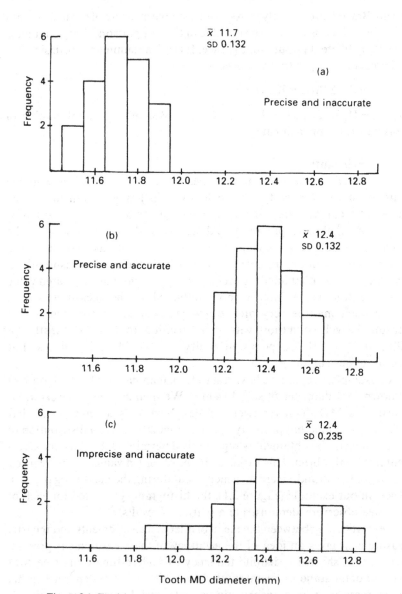

Figure 2.1. Precision and accuracy as reflected in the histograms of the frequency of measurements of a phantom tooth by three observers – (a), (b) and (c).

determined values bunch about the mean. In our example, observers (a) and (b) maintained high levels of precision (SD = 0.132) whilst observer (c), who demonstrated a wide scatter of results, was imprecise (SD = 0.235). Hence, we may say that observer (a) was precise but inaccurate, observer (b) was both precise and accurate and observer (c) was accurate but imprecise.

From the foregoing example, we may conclude that reliability is not only concerned with either precision (repeatability) or accuracy (bias), but with both of these factors. We may also say that each measurement we take is made up of two parts, a true component and an error component. Hence:

$$X = X_t + X_e$$

where    $X$ = any measurement
          $X_t$ = true component
          $X_e$ = error component.

The error component has two elements – a random element, which relates to precision (repeatability) and a systematic element which relates to the accuracy (bias) of measurement. The magnitude of the error component is of course related to the reliability of the measuring method, but it cannot provide information about its nature. That is, it cannot identify which source of variability has the greater influence on reliability. Hence these two sources of error (systematic or random) need to be investigated separately.

## 2.4    Sources of error

### 2.4.1    *Impression and casting techniques*

Elastic impression materials have improved in accuracy and ease of manipulation since the introduction of agar-based reversible hydrocolloid in 1929. Modern alginates (irreversible hydrocolloids) are as accurate as reversible hydrocolloids and are superior in the physical properties of strain under compression, set in compression and strain at the point of rupture (Davis & Preble, 1986). The accuracy of a model cast from an impression will obviously relate to the dimensional stability of the impression and of the cast. Since both these factors may directly be affected by operational technique, it is imperative that the operator should avoid those inaccuracies that are under his control. For instance, attention should be paid to the correction of syneresis (shrinkage due to water loss) and imbibition (expansion due to water absorption), the correct tray filling technique should be used, the prescribed mixing time should be adhered to, and finally, casts should be poured as soon as possible after removal of the impression from the mouth.

Table 2.1. *Age-related changes in the lengths of interproximal contact areas (mm) as a measure of interstitial attrition in the Lengua Indians of Paraguay (Kieser et al., 1985b)*

| Age group (y) | P1 | | | P2 | | | M1 | | | M2 | | | M3 | | |
|---|---|---|---|---|---|---|---|---|---|---|---|---|---|---|---|
| | $n$ | $\bar{X}$ | SD | $n$ | $\bar{X}$ | SD | $n$ | $\bar{X}$ | SD | $n$ | $\bar{X}$ | SD | $n$ | $\bar{X}$ | SD |
| *Maxilla* | | | | | | | | | | | | | | | |
| -19 | 32 | 2.75 | 1.4 | 34 | 2.72 | 0.6 | 37 | 3.65 | 1.2 | 37 | 4.07 | 1.2 | 11 | 4.52 | 1.7 |
| 20-29 | 74 | 2.71 | 1.0 | 76 | 3.67 | 1.5 | 69 | 4.90 | 1.2 | 69 | 5.32 | 1.2 | 38 | 4.95 | 1.3 |
| 30-39 | 46 | 3.35 | 1.0 | 50 | 4.19 | 1.0 | 49 | 5.46 | 1.1 | 45 | 6.07 | 1.0 | 30 | 5.37 | 0.8 |
| 40-49 | 21 | 3.30 | 0.6 | 24 | 4.23 | 1.6 | 26 | 5.38 | 1.8 | 22 | 6.00 | 1.4 | 24 | 5.68 | 1.1 |
| 50+ | 25 | 3.81 | 0.6 | 22 | 4.90 | 0.6 | 22 | 6.28 | 0.9 | 17 | 5.97 | 1.0 | 9 | 6.07 | 1.2 |
| *Mandible* | | | | | | | | | | | | | | | |
| -19 | 33 | 2.55 | 1.4 | 34 | 2.62 | 0.8 | 35 | 3.17 | 1.0 | 36 | 3.75 | 1.0 | 32 | 3.75 | 1.2 |
| 20-29 | 66 | 2.69 | 0.8 | 74 | 3.02 | 0.8 | 68 | 3.96 | 1.0 | 67 | 4.25 | 1.1 | 34 | 4.35 | 1.7 |
| 30-39 | 38 | 3.01 | 0.9 | 44 | 3.69 | 1.0 | 39 | 4.84 | 1.0 | 35 | 5.18 | 1.2 | 25 | 4.82 | 1.3 |
| 40-49 | 23 | 2.79 | 0.7 | 28 | 3.48 | 1.0 | 25 | 4.68 | 0.9 | 26 | 5.48 | 1.2 | 18 | 4.50 | 0.9 |
| 50+ | 17 | 2.77 | 0.5 | 15 | 3.75 | 0.7 | 26 | 4.88 | 0.8 | 24 | 5.64 | 1.0 | 15 | 5.20 | 1.1 |

Although it may be accepted that careful technique will assure greater precision, the results of laboratory investigations imply that the effects of material distortion and shrinkage remain unpredictable and not readily controlled by the operator (Hollinger *et al.*, 1984). Hence, regardless of the technique or material used, some linear distortion has to be accepted. Hollinger *et al.* (1984) documented linear differences from the standard of $-0.38\%$ to $0.82\%$ in mandibular impressions and $-0.09\%$ to $0.56\%$ in maxillary impressions. This means that one would have to accept a linear discrepancy of around 0.04 mm for the average sized mandibular molar $(MD = 11\,mm)$ and of 0.06 mm for lower molars.

## 2.4.2    Attrition

Central to an appreciation of dental attrition as a source of measurement error is an understanding of the purpose of the study under consideration. Whilst to the geneticist, unworn teeth would be ideal, the orthodontist might consider worn and unworn teeth to be of equal importance. Those interested in human adaptation, on the other hand, might require a comparative evaluation of worn and unworn teeth.

Attrition may be defined as the frictional wearing away of the substance of a tooth under the influence of mastication (Moorrees, 1957; Molnar, 1971). Wear may be either occlusal, or interproximal, or a combination of both. Interproximal attrition has been shown to result in a progressive increase in the interproximal contact areas of adjacent teeth. This was

Table 2.2. *Tooth size reduction due to attrition in the San (Bushmen).* (*Computed from figures published by van Reenen, 1982. All measurements in mm.*)

| Tooth | Dimension | Maxilla | | Mandible | | $\bar{x}$ |
|---|---|---|---|---|---|---|
| | | Male | Female | Male | Female | |
| I1 | MD | 2.6 | 1.1 | 1.5 | 1.4 | 1.6 |
| | BL | 0.6 | 0.2 | 0.1 | 0.3 | 0.3 |
| I2 | MD | 1.4 | 1.5 | 1.7 | 1.3 | 1.4 |
| | BL | −0.1 | 0.8 | −0.3 | 0.0 | 0.5 |
| C | MD | 0.9 | 0.7 | 0.9 | 1.0 | 0.9 |
| | BL | −0.2 | 0.1 | −0.6 | −0.4 | −0.3 |
| P1 | MD | 0.7 | 1.3 | 1.5 | 1.1 | 1.1 |
| | BL | 0.0 | −0.1 | 0.2 | −0.1 | 0.0 |
| P2 | MD | 1.1 | 0.6 | 1.6 | 1.9 | 1.3 |
| | BL | 0.1 | 0.2 | 0.3 | 0.2 | 0.2 |
| M1 | MD | 1.7 | 1.5 | 2.2 | 1.5 | 2.0 |
| | BL | 0.3 | 0.8 | 0.3 | 0.6 | 0.5 |
| M2 | MD | 1.2 | 1.2 | 0.4 | 0.7 | 0.9 |
| | BL | 0.3 | −0.3 | 0.2 | — | 0.2 |

exemplified by a recent evaluation of dental wear in the preliterate Lengua Indians of Paraguay (Kieser, Groeneveld & Preston, 1985b, c). When these authors contrasted age related changes in interproximal contact lengths among the postcanine teeth (Table 2.1), they found that attrition resulted in a mean increase of 1.14 mm in premolars and of 1.85 mm in molars. The consequent shortening of the MD diameter due to interproximal attrition was highlighted by van Reenen (1964), in his study of another preliterate population, the Kalahari San (Bushmen).

Table 2.2 lists the dimensional differences between unworn and heavily worn teeth in the San. The range of mean reduction in the MD diameter is 0.9 to 2.0 mm, with the greatest dimensional changes being found in first molars and first incisors. Least affected were canines and second molars. In the BL dimension, mean changes ranged from −0.3 to 0.5 mm, suggesting that some teeth (notably the canines), actually increased in size. This may be explained by the fact that the points of maximum convexity of tooth crowns are often well gingival to the occlusal surface. Hence, unless maximum measurements are taken, occlusal wear will result in an apparent dimensional increase.

Clearly attrition may have an appreciable effect on the diameters of teeth. There are those who have suggested that where attritional excoriation was slight, and the amount lost could be estimated with a reasonable level of accuracy, the MD diameter could be corrected so that it approximated the unworn diameter (Doran & Freedman, 1974; Grine, 1981). Others, however, have suggested that only teeth without marked attrition should be used for odontometric purposes (Potter *et al.*, 1981; Kieser, Preston & Evans, 1983; Kieser, 1985; Kieser *et al.*, 1985a, b, c).

### 2.4.3    Dental calculus

There is a curious, though well documented, independence between the MD and BL diameters of human anterior teeth (Garn, Lewis & Kerewsky, 1968b; Potter *et al.*, 1968, 1976; Lombardi, 1975; Townsend & Brown, 1979a, b; Kieser *et al.*, 1985a, b). The interpretation of this finding has tended to be in terms of 'independent genetic determinants' or 'overlapping morphogenetic effects' (see Kolakowski & Bailit, 1981a). However, it has been suggested that a rather prosaic factor, namely dental calculus, might be responsible (Kolakowski & Bailit, 1981a).

Dental calculus is hard, mineralised, accumulation of variable thickness which may form above or below the gingivae. Grant, Stern & Everett (1979) have noted that supragingival calculus tends to be highly irregular and may add up to 5 mm to a tooth's buccolingual diameter. Although most odontometric studies utilise measurements taken on dental casts, reliable evaluation of calculus can only be made intraorally. It is for this reason that a number of investigators have chosen to omit BL measurements of upper and lower incisors unless there are no calculus deposits on these teeth.

### 2.4.4    Other factors

Other sources of error in the metrical evaluation of human teeth relate to the loss of tooth substance, either due to caries or mechanical intervention by dentists. Although dental decay should be obvious enough, even to the nonclinician, well-contoured restorations and carved cast crowns may present a problem when measurements are to be taken from casts. Hence measurements should not be attempted on carious, restored or fractured teeth.

## 2.5    Error measurement

### 2.5.1    The coefficient of reliability

One of the most widely used error measurements is the reliability coefficient; teeth are measured and then remeasured at a later date. The investigator then computes a Pearsonian product–moment correlation coefficient between the two sets of measurements. Sadly, the reliability coefficient has recently been shown to be both inappropriate and inadequate (Bland & Altman, 1986; Day & Keyser, 1987). Their reasons were as follows: first, the correlation coefficient measures the strength of a relation between two variables and not the agreement between them. Secondly, the correlation depends on the range of the measurements in the sample. That is, the wider the spectrum of measurements, the greater the correlation. Since researchers usually compare the two sets of measurements over the whole range of tooth measurements typically encountered, a high correlation is almost inevitable. Thirdly, two sets of measurements of the same entities are by definition related, and the test of significance ($H_0$ = Zero correlation) will reflect this. Hence the test of significance is irrelevant to the question of agreement. Finally, data which seem in poor agreement, have been shown to produce highly significant correlations (Day & Keyser, 1987). For these reasons, it should be clear that the fact that two sets of measurements are correlated is of little significance in assessing error measurement. What needs to be measured is the extent to which the two sets differ. One method of measuring this was described by Bland & Altman (1986) and involves simply plotting the difference between each pair of measurements against their mean. The plot then allows any possible relationships between the true component and the error component to be investigated.

## 2.5.2 Standard error of measurement
Rather than use correlation coefficients to express the reliability of a set of measurements, one may use the standard error of measurement. This is not affected by the range of values of the sample tested. The standard error of measurement is the standard deviation of a sample of an individual's scores about the true score. Suppose one administered the same set of measurements repeatedly and then took the mean of them all. One could use this as an estimate of the true score with the standard deviation of these measurements referred to as the standard error of measurement. The smaller the standard error of measurement, the more reliable the measurement technique. In fact, a perfectly reliable instrument and technique would have a zero standard error of measurement.

## 2.5.3 Method error statistic
Measurement precision may also be evaluated by means of the method error statistic. This statistic represents an attempt to quantify the inherent imprecision of a single observation of a variable as determined by duplicate measurements of that variable (Dahlberg, 1940; Solow, 1966; Utermohle, 1983; Utermohle, Zegura & Heathcote, 1983). This method has been employed principally by orthodontists for cephalometric variables (e.g. Bjork, 1947; Solow, 1966; Krogstad, 1972; Haugen, 1977). This method has also found favour with anthropometrists, who refer to it as the

technical error measurement (e.g. Johnston, 1980; Branson *et al.*, 1982; Cameron, 1984). The method error statistic ($S$) is calculated as follows:

$$S = \sqrt{\left( \Sigma \frac{(x_1 - x_2)^2}{2n} \right)}$$

where $x_1$ is the original measurement value, $x_2$ is the repeated measurement value and $n$ is the sample size. Those who favour this statistic have often assumed that the distributions of the replicate difference are normal and that the errors of all pairs of measurements can be pooled (see Utermohle *et al.*, 1983). There is considerable controversy around this assumed normality of the distribution of replicate differences, in spite of which the statistic remains a useful one, partly because by employing squared differences, it weights imprecision according to the magnitude of the error.

### 2.5.4    Student's t-test

Earlier it was noted that the relationship between the mean of a set of measurements and the mean of a repeated set of measurements gives an idea of the bias of the measurement process. Differences between the means of two sets of measurements may easily be evaluated by means of Student's *t*-test. It is worth remembering, however, that the *t*-test is based upon mean differences and is thus primarily an indicator of systematic and not random error.

### 2.5.5    Analysis of variance

The reliability of a measurement technique expresses the ratio of true variance to true variance plus error variance (Buschang, Tanguay & Demirjian, 1987). As we have seen $X = X_t + X_e$, where $X$ is any measurement, $X_t$ is the true component of variance and $X_e$ is the error component, consisting of a random and a systematic variance component. These may be evaluated by an array of different univariate statistics. However, the situation is often encountered where there are different observers using different instruments. The very complexity of intra- and interobserver error combined with instrumental and temporal error requires a multivariate approach such as the analysis of variance. Replicate measurements by more than one observer may be evaluated by means of analysis of variance examining the effects of observers, instruments, age, sex, etc., as well as the interactions between and among them. Of course, the design becomes more complex with each additional source of variation. A full analysis allows for the assessment of global reliability for multiple operators and different types of teeth (Buschang *et al.*, 1987).

# 3   *Inheritance of tooth size*

## 3.1   Introduction

Although modern dental genetics had its roots in Aristotelian science, it really owes its origin to the work of Gregory Bateson and to the rediscovery in 1900 of Gregor Mendel's work. Bateson introduced the concept of teeth as serially repeated or meristic structures (Bateson, 1894) which, he suggested, should be viewed as a unit, varying and evolving as a whole. Bateson further characterised the properties of a meristic series as follows: first, members of a series show serial homology and hence differences between them should be seen as quantitative rather than qualitative; second, the form of each member of a series is dependent upon its position within the series as a whole. Three of Bateson's contemporaries, however, held opposing views, regarding units within the dentition as functional rather than morphogenetic. Scott (1892) suggested that the apparent serial homology between premolar and molar cusps was nothing more than the deceptive result of convergent evolution, a view which implied that each tooth was under independent hereditary control. Rutimeyer (1863) and Wortman (1886) suggested that all the teeth were in fact modified molars, with the implication that all the teeth shared the same hereditary determinants. It was when the importance of morphological variation began to be realised, toward the close of the nineteenth century, that questions concerning its causation began to be asked. The foremost question concerning inheritance asked during this period was whether the totality of characters was controlled by a single substance or by separate, independently varying particles.

Darwin's half-cousin, Francis Galton, became similarly concerned with generalised nondiscrete characters and proposed his own theory of inheritance, which saw the individual as the sum total of genetic particles, some of which remain dormant, whilst comparatively few achieve expression (Galton, 1876).

As Mayr (1982) noted, it remains one of the great ironies in the history of science that the answer to the problem of heredity had already been found while so many investigators searched for it so diligently. The three botanists – de Vries, Correns and Tschermak – who rediscovered Mendel's genetic theory, introduced into the biological world a theory of inheritance which was based on a single-particle postulate, and with it they introduced the often misconceived notions of genotype and phenotype. The failure to make a clear distinction between an underlying genetic factor and the resultant phenotype, together with the failure to reconcile discontinuous Mendelian inheritance with continuous variation, was to puzzle dental geneticists for many years.

## 3.2    Teeth, nature and nurture

Francis Galton remarked, over a hundred years ago, that: 'Twins have a special claim upon our attention, it is that their history affords a means of distinguishing between the effects of tendencies received at birth and of those that were imposed by the special circumstances of their after-lives' (Galton, 1883).

It was more than four decades before Galton's call was answered. In 1927 Bachrach and Young wrote a now almost forgotten article in which they used twins to obtain a measure of the relative roles of 'heredity' and 'environment' in producing observed variation in tooth eruption times, dental caries, enamel hypoplasia and normal occlusion. Goldberg (1929, 1939) extended Bachrach and Young's investigation on malocclusion and concluded that heredity played a large role in its development. About the same time Korkhaus (1930) studied odontometric variability of the anterior teeth in twins and found that the largest intrapair differences in size occurred in the maxillary lateral incisors. Cohen and his co-workers studied nine sets of triplets in whom they recorded the dimensions of the teeth and they concluded that tooth size was not highly heritable (Cohen, Oliver & Bernick, 1942). Tobias (1955) followed with an analysis of fraternal and identical twins from which he concluded that dental differences which had developed between identical twins must be attributed to the differential operation of environmental factors after the splitting of the embryonic axis.

Lundström (1948, 1954, 1955, 1963, 1967) furthered the study of the dental features of monozygotic and dizygotic twins by measuring genetic and nongenetic variance ratios. Although the major emphasis of his work was on dental arcade and profile relations, Lundström was able to show that dental dimensions may be successfully employed in the diagnosis of zygosity in twins. Like Lundström, Horowitz and his colleagues investigated hereditary variation in the MD diameters of anterior teeth (Horowitz, Osborne & de George, 1958), and confirmed the former's results: the greatest variation was found in the lateral incisors, whilst the canines demonstrated the least measurable genetic variability. These results were further confirmed in a subsequent study of the dental dimensions of triplets (Menezes, Foster & Lavelle, 1974).

Hence much of the impetus for the study of dental features in multiple births came from a desire to assign these individuals to the correct class of zygosity. Then came great emphasis on using data from twins to interpret the complicated phenomena surrounding inheritance, particularly after Bulmer (1970) and Smith (1974, 1975) proposed a formula from which to estimate heritability from this data:

$$h^2 = 2(R_{mz} - R_{dz})$$

where $R_{mz}$ and $R_{dz}$ are the respective intraclass correlation coefficients of monozygous and dizygous twin pairs. Utilising this formula, Potter & Nance (1976) and Mizoguchi (1977) showed that relatively low heritability of MD tooth size characterised the upper canines and first premolars and also the mandibular lateral incisors, the premolars and the first molars. These teeth were those that had been shown to be the most stable during human evolution. Noting Falconer's (1967) suggestion that characters with low heritabilities tended to be closely connected with reproductive fitness, Mizoguchi (1977) was tempted to suggest that the more stable teeth in man were controlled by a shared genotype.

Prompted by the evidence from previous twin studies for the existence of genetic control of tooth size, Potter and her co-workers questioned whether each dental measure was under independent genetic control, or whether dental measurements were influenced by a number of common genetic determinants (Potter *et al.*, 1976). In order to demonstrate the presence of independent genetic determinants for the multiple correlated dimensions of teeth from twin data, these investigators applied multivariate analysis to the covariance matrices of monozygotic and dizygotic within-pair differences for both MD and BL dimensions. Their results provided strong evidence for a pleiotropic effect of independent genes or of groups of genes. Interestingly, they found that genetic determination of the maxillary and mandibular teeth were independent of each other.

Twin studies having clearly shown that hereditary factors act during odontogenesis to control tooth size at least partially, researchers now turned to family studies. Earlier, Garn, Lewis & Kerewsky (1965a, c) had used paired sibling mean product–moment correlations for each of the permanent teeth to show that X-linkage was one hereditary mechanism which was partially responsible for the control of MD tooth size. These investigators found that sister–sister correlations were higher than brother--brother correlations which in turn, were higher than brother–sister correlations. These results would, of course, be expected with X-linked inheritance, since sisters always share a common paternal X chromosome, but, along with brother–sister pairs, have an equal chance of sharing the same maternal X chromosome (Mather & Jinks, 1963). Further support for an X-linked component in inheritance of tooth size came four years later when Lewis & Grainger (1967) examined tooth size correlations for homologous teeth in seven groups of parent–child and sibling pairs. The correlations were, however, too low for use in the prediction of tooth size. Potter *et al.* (1968) criticised the use of twin studies, asserting that the most careful investigations would still only yield the broad conclusion that persons who share a common genotype are more likely to react similarly to their environment than those persons with dissimilar genotypes. Citing Falconer (1967), these authors noted that the genotypic variance obtained was composed of additive, dominant and epistatic components, the additive genetic variance being attributable to the effects of allelic genes taken singly and summed over all the loci. This prompted Potter and her co-authors to argue that since parents transmit genes and not genotypes to their offspring, additive inheritance (which refers to genes and not genotypes) should be the principal focus of dental genetic investigations (Potter *et al.*, 1968). However, when they used factor analysis on the covariances of 635 individuals from 153 Pima Indian families, they found that although the total genetic variability was detectable, its components (i.e. additive, autosomal or dominance effects) were not. They concluded that environmental factors must have been responsible for the inability to discern the specific genetic components of variance from their data set.

In Britain, Goose (1967, 1971) calculated heritabilities of MD diameters of maxillary incisors and canines in the members of 123 families living in the Liverpool area. High heritabilities and high parent–offspring correlations indicated little environmental influence and were compatible with a hypothesis of multifactorial inheritance with genes of small and additive effect, but without dominance. Unlike Garn *et al.* (1965a, c), Goose was unable to demonstrate sex-linkage in the inheritance of tooth size.

To separate nongenetic from genetic tooth size determinants, Alvesalo & Tigerstedt (1974) used heritability estimates from 90 groups of full siblings. Their results showed a clear range of variation of heritabilities, distal members of each tooth group showing the lowest heritability estimates. This finding suggested to the authors that a lower heritability meant that the tooth in question would be less liable to evolutionary change.

The role of sex chromosomes in the inheritance of tooth size was again raised by Townsend & Brown (1978a), who calculated product–moment correlation coefficients between different full-sibling and half-sibling pairs of Yuendumu Aboriginals living in the Northern Territory of Australia. Their lack of evidence for sex chromosomal involvement was at variance with the findings of Garn *et al.* (1965a, c).

Townsend & Brown (1978b) further argued that heritabilities derived from twin or full-sibling data tended to be elevated because of shared environmental effects. On the other hand, paternal half-sibling data are less likely to be influenced by a shared environment and consequently provide more reliable estimates of heritabilities. The custom of polygyny practised by the Yuendumu Aboriginals provided these authors with a rare opportunity for analysing data derived from groups of full-siblings and half-siblings.

Their results did not agree with those of Garn *et al.* (1965b) that genetic factors may account for up to 90% of the total variability in tooth size. Instead heritability values derived from their data suggested that the contribution of additive genetic variance was less in the Australian Aboriginals studied – around 64% on average. The variance due to common environmental factors was estimated to account for about 6% of the total variance of permanent tooth size. Interestingly, these authors found some evidence that the common environmental influence on BL dimensions exceeded that of the MD dimensions.

### 3.3 Unifactorial and multifactorial inheritance

This last group of investigations (Garn *et al.*, 1965a, b, c; Lewis & Grainger, 1967; Goose, 1967, 1971; Potter *et al.*, 1968; Alvesalo & Tigerstedt, 1974; Townsend & Brown, 1978a, b) was made possible by another principal advance in genetic thinking earlier in the century. With recognition of the distinction between the genotype and the phenotype (Johannsen, 1909) came the realisation that intermediate forms, previously attributed to 'blending inheritance', could not be accounted for by the inexorable genetic determinism of Mendel's laws. The Swedish plant breeder, Nilsson-Ehle, first demonstrated experimentally that quantitative characters, showing continuous phenotypic variation, may be inherited in a strictly Mendelian fashion. The theory that multifactorial inheritance converted discontinuous genetic variation into continuous phenotypic variation was subsequently used by the Morgan School to argue against the one-gene–one-character theory of early Mendelism (Mayr, 1982). Beginning with Bateson's (1913) concept of epistasis, which referred to interactions among different gene loci, and Chetverikov's (1926) concept of a genetic milieu contributed to by all the genes of an individual, geneticists began to analyse the continuous variation documented by the bio-metricians. The new science of biometric genetics was formalised by Fisher in his famous series of papers of 1918 and 1922, which dealt with the mathematics of gene distributions in populations. Fisher partitioned the genetic variance into an additive portion, which he suggested was caused by alleles or independent genes with similar effects, and a nonadditive portion, which referred to epistasis and dominance. Fisher's emphasis on genes with small phenotypic effects led him to conclude that continuous variation was due to multiple Mendelian factors rather than to environmental factors.

Another geneticist, Sewall Wright, disagreed with Fisher on this point and so revived the old nature–nurture argument which has generated immense scientific interest to this day. In his analysis of IQ in adopted children, Wright (1931) postulated the existence of direct inheritance of 'environmental factors', a concept which was supported over forty years later by the proposal of Cavalli-Sforza & Feldman (1973) that another mode of inheritance exists (referred to by them as cultural inheritance) which operates side by side with strictly biological (DNA-based) inheritance.

Genetic variation underlies the expression of such diverse human traits as allozymes, polydactyly, the blood groups, stature, albinism and tooth size. Some traits have a relatively simple pattern of inheritance because they are affected by relatively few genes, for example the A, B, O blood groups which are controlled by the alleles at a single locus. On the other hand, the inheritance of those differences between individuals that are of degree rather than of kind, depends on alleles at several loci. Such traits, of which tooth size is an example, are said to be polygenic or multifactorial. The distinction between unifactorial (or Mendelian) inheritance and multifactorial (or polygenic) inheritance lies not in genes with fundamentally different properties, but rather in the number of genes and the magnitude of the genetic effect relative to other sources of variation. However, the methods of study employed in Mendelian genetics differ from those used in polygenic or quantitative genetics. First, the unit of study has to be expanded from single progenies to larger groups of individuals or populations comprising many progenies. Secondly, polygenic traits are usually also influenced by the environment to varying degrees. This means that variation in tooth size, for example, will be due partly to genetic variation among individuals and partly to environmental factors, both of which will have to be considered in an evaluation of heritability.

Differences in degree rather than in kind, obviously involve variation without natural discontinuities. This is called continuous variation of a range of phenotypes and such characters are referred to as quantitative or metric characters. For ease of reference, however, one might consider grouping phenotypes arbitrarily into categories such as those illustrated in Figure 3.1. Here the variation in premolar size in males is represented graphically by a frequency distribution of measurements, with the proportion of individuals within equally spaced classes plotted on the vertical scale. In this case, men with premolar diameters between 6.45 and 6.55 mm, are treated as if each had a crown size of 6.5 mm. This grouping produces a set of measurements (phenotypic values) denoted $x_1, x_2, \ldots, x_n$ with respective frequencies of $p_1, p_2, \ldots, p_n$.

Importantly, the resulting histogram is discontinuous only for the sake of

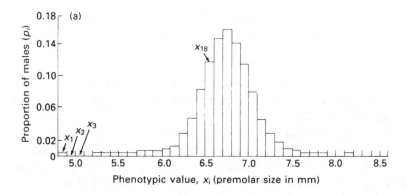

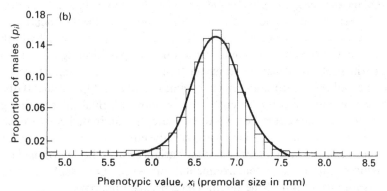

Figure 3.1. (a) A hypothetical distribution of premolar size among a number of males. MD classes are denoted $x_i$, the proportion of men in each class being denoted by $p_i$. (b) A normal curve but approximating the MD distribution in (a). (Adapted from Harrison *et al.*, 1964.)

convenience of plotting. If the ranges of individual classes could be made infinitely small, the histogram would of course become a smooth curve, with a more or less normal distribution. Thus it becomes possible to make use of the properties of the normal distribution in analyses of polygenic traits. For the hypothetical data in Figure 3.1 for example, $p_i$ is the proportion of individuals that have a phenotypic value of $x_i$. It follows that $p_1 + p_2 + \ldots + p_n = 1$, and that the mean of $x$ is:

$$\bar{x} = \frac{p_1 x_1 + p_2 x_2 + \ldots + p_n x_n}{n}.$$

Equally important in summarising quantitative traits is the variance ($\sigma_x^2$), which measures how closely individual observations cluster about the mean. The variance of $x$ is calculated by working out the deviation of each value from the population mean $(x_1 - \bar{x}; x_2 - \bar{x}; \ldots; x_n - \bar{x})$, squaring each of these deviations and then taking the average:

$$\sigma_x^2 = p_1(x_1 - \bar{x})^2 + p_2(x_2 - \bar{x})^2 + \ldots + p_n(x_n - \bar{x})^2.$$

The value of the normal distribution in quantitative genetics now becomes clear: the entire distribution of such traits may be conveyed by two figures, the mean and the variance.

The amount of variation as measured and expressed by the variance is extremely useful in determining the genetics of metric characters. Basically, one wishes to partition the components of variation into genetic and environmental contributions. Whilst the genetic portion refers to the assemblage of genes possessed by the individual, the environmental portion will refer to all the nongenetic factors which might influence the phenotype. In our example, the variance in premolar size refers to that caused by both the genetic and the environmental variation. This partitioning thus refers back to the problem of 'nature versus nurture', or the relative importance of genotype and the environment in determining the phenotypic value. If one of these components could be completely eliminated, then the remaining phenotypic variance could be used to estimate the eliminated component. Because it contains all nongenetic variance, the environmental component cannot be removed, and hence genotypic variance has to be determined on the assumption of equal environmental influence. One such method involves an evaluation of the resemblance between relatives.

## 3.4    Heritability

The degree of resemblance between relatives (which may of course be determined by simple measurement) provides a useful means of estimating the amount of genetic variance. Thus one can say that the relative importance of the genotype as a determinant of the phenotypic value may be given by the ratio of the genetic variance to the phenotypic variance:

$$h^2 = \frac{\sigma_g^2}{\sigma_p^2}.$$

This is referred to as broad-sense heritability. However, quantitative geneticists often use the word heritability in reference to a different variance ratio, namely the ratio of the additive genetic variance to the total phenotypic variance. The total genetic variance used earlier is in fact the additive variance plus the epistatic or interaction variance, plus the effects of dominance. Consideration of these factors are outside the scope of the present discussion. However, it is important to remember that the components of genotypic variance represent the cumulative effect of all loci affecting the trait.

The theoretical formula for heritability says virtually nothing about the actual mode of inheritance of a quantitative trait. It merely expresses the proportion of the total variance that may be attributed to the average effects of genes. Heritability is a property not only of a character but also of the population and of the environmental circumstances to which individuals in the population are subjected (Falconer, 1967). Since the magnitude of the heritability depends on the magnitude of all the components of variance, a change in any one of these may affect it. Hence the value stated for a given character can only refer to a particular population, living under particular environmental conditions.

## 3.5    The estimation of heritability

In human studies, estimates of heritability have to be derived from measuring the similarity between relatives. There are many degrees of relationship (e.g. parent–offspring, sibling, half-sibling, first cousin) that may be relied on and the best estimates probably derive from the comparison and combination of a number of these, but only one of these will be briefly discussed here: the relationship between parents and their offspring.

Mean measurements for each tooth class in a number of sons ($y$-values) may, for instance, be plotted against the phenotypic values of their father ($x$-values). A line of best fit applied to such a bivariate plot will thus give a regression line of offspring on parent. Once again, it is the variance that is important: this time, however, one needs to consider how the two sets of measurements covary. The covariance of $x$ and $y$ ($\sigma_{xy}$) is the average of the product of the deviations of $x$ and $y$ from their respective means. Hence, if $\sigma_{xy}$ represents the covariance between tooth size values of fathers and those of their male offspring, and $\sigma_x^2$ represents the variance of phenotypic values of the father (recall Figure 3.1), then the slope of the regression of $x$ on $y$ is simply $\sigma_{xy}/\sigma_x^2$. The slope ($b$) is called the regression coefficient of offspring on one parent and enables the heritability to be estimated from the relationship $b = \frac{1}{2}h^2$ (divided by two because the regression involves only a single parent). Generally speaking, the regression of offspring on father is more informative than the regression of offspring on mother, because the mother's contribution includes more environmental influence.

One attempt at elucidating the problem of tooth size determination by using heritability estimates as an indicator of genetic effects was that of Alvesalo & Tigerstedt (1974). Using full-sib analyses and broad-sense estimates of heritabilities, these authors showed that genetic variance appeared to be higher on mesial members of tooth groups than on distal members. Notable exceptions to this general rule were the maxillary premolars (MD dimension) and the mandibular second incisors. The authors also stressed that the relative phenotypic variability of distal teeth was higher than that of mesial members of the same tooth group, with the sole exception of the lower incisors. This, in turn, was interpreted as signifying that the lower the total genetic or additive portion of tooth size variance, the greater the phenotypic variability in size. Whilst it was clear that their results provided at least some support for Butler's field concept (Butler, 1939), the investigators remained perturbed about some of their findings. In an attempt to account for the observed inverse relationships, they uncovered an apparent contradiction which has, as yet, not been adequately explained. If it is accepted that tooth size and individual selective fitness are related and that the rate of increase in fitness is equal to the additive genetic variance (Fisher, 1930), then it is difficult to explain Falconer's (1967) observation that characters more closely connected with fitness have lower heritabilities. Alvesalo & Tigerstedt (1974) felt that teeth with higher heritabilities would be favoured by factors which would tend to maintain developmental stability. Hence, one would expect the polar teeth within each tooth class to have the lowest heritabilities. However, hominid facial reduction placed an entirely different perspective on the selective fitness of teeth. In the face of steadily reducing jaw size, it was reasoned, more variable teeth would be selectively advantageous because they could respond more readily to spatial restriction. Hence it had become intra-arcadal harmony rather than odontometric stability which was related to fitness. Support for this hypothesis has come from the genetic analyses of Mizoguchi (1977) and Harzer (1987) and from the theoretical model proposed by Sofaer, Bailit & MacLean (1971) (Chapter 8).

## 3.6    Gene effects

Genetic involvement in human tooth size is usually demonstrated by a nonzero heritability. The question remains, however, of what genetic mechanisms underlie individual and population differences in tooth size. Kolakowski & Bailit (1981b) attempted to identify specific loci which determine significant amounts of odontometric variation in Solomon Islanders. A large sample (3000 individuals) from nine language groups, was subjected to the segregation analysis developed by Elston & Stewart (1971). This statistical method combines the advantages of a maximum likelihood procedure with the utilisation of multigenerational pedigrees.

Despite the large sample size and their sophisticated technique, their results were ambiguous. When they limited themselves to the results of one diameter in isolation, they were able to refute the hypothesis of a major locus effect. When only nuclear families were considered, their model suggested that a polygenic explanation fitted equally well. Invoking the principle of parsimony, these authors concluded that no major locus effect should be asserted for human tooth crown size.

## 3.7 Nonrandom environmental effects

According to the classic model for the multifactorial inheritance of quantitative traits, phenotypic similarity between relatives is due to genetic and biological environmental factors. Research in the last decade or so has modified this model substantially. Attention has been drawn to the importance of nonrandom environmental factors in the transmission of quantitative characters. For example, Cavalli-Sforza & Feldman (1973) introduced a model in which cultural and biological inheritance were both determined by a single locus. More recent research has shown that although polygenic traits may be highly familial, it is necessary to allow for genetic and cultural transmission, as well as for systematic effect due to a shared rearing environment (Rao, Morton & Yee, 1976; Rice, Cloninger & Reich, 1978, 1980). These suggestions have thus gradually shifted the emphasis in the determination of family resemblance from purely genetic factors to nonrandom and nonbiological environmental factors. Hence, for the evaluation of dental genetics, it has become necessary to adopt a model which does not exclude *a priori* either genetic or cultural transmission. One such model is the general linear model of familial resemblance originally developed by Morton (1974) and Rao and his associates (1974, 1976). More recent modifications have expanded it into a useful method which accommodates polygenic inheritance, cultural transmission, phenotypic assortative mating and a common rearing environment (Rice *et al.*, 1980). What follows is a brief description of this path analytic model, without consideration of its complex mathematical background for which reference may be made to the works of Rice and his co-workers (1978, 1980, 1981), Cloninger, Rice & Reich (1979a, b) and Rao *et al.* (1974, 1976).

Essentially, path analysis assumes that a given phenotypic end-product ($P$) will be derived as follows: $P = T + E$, where $T$ denotes both cultural and genetic factors transmissible from parent to offspring and $E$ denotes all other effects. If one now considers a quantitative phenotype such as tooth size, then one may partition the distributions of tooth size for males ($P_1$) and females ($P_2$) as:

$$P_1 = \alpha_1 + T_1 + E_1 \text{ and } P_2 = \alpha_2 + T_2 + E_2.$$

Here it is assumed that $P$ is measured on a scale where males have a mean of $\alpha_1$ and females have a mean $\alpha_2$.

Now let $T_i$ denote the effects of all factors relevant to the development of

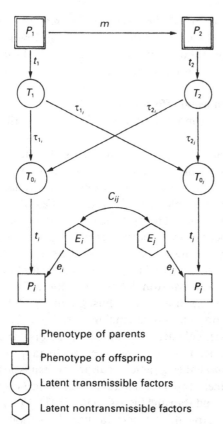

▨ Phenotype of parents

□ Phenotype of offspring

◯ Latent transmissible factors

⬡ Latent nontransmissible factors

Figure 3.2. A path diagram depicting the sources of resemblance between parent and offspring and between two siblings in the presence of sex dimorphism. Observed quantities are in boxes, $P$ denotes phenotype, $T$ is genetic and cultural factors that are transmissible from parent to offspring and $E$ denotes non-transmissible factors. (Adapted from Rice *et al.*, 1980. See text for details.)

tooth size $P_i$ which are transmissible from a parent of sex $i$ to an offspring of either sex (Figure 3.2). The effects of all other factors relevant to the development of $P_i$ (and which are nontransmissible) are denoted by $E_i$, while $C_{ij}$ represents such factors common to both sexes. If the variables are standardised, the following path equations result:

$$P_1 = t_1 T_1 + e_1 E_1 \text{ and } P_2 = t_2 T_2 + e_2 E_2$$

where the path coefficients $t_i$ and $e_i$ are given by:

$$t_i = \sigma_{T_i}/\sigma_{P_i} \text{ and } e_i = \sigma_{E_i}/\sigma_{P_i}.$$

Family resemblance ($T_{0_i}$) is determined by the path equation

$$T_{0_i} = \tau_{1_i} T_1 + \tau_{2_i} T_2 + e_i E_i$$

where $T_1$ and $T_2$ denote latent transmissible factors bequeathed by the father and mother respectively. The path coefficients $\tau_{i_j}$ are standardised partial regression coefficients which measure the contribution of a parent of sex $i$ to an offspring of sex $j$ (Rice *et al.*, 1980).

Having run briefly through the mechanics of path analysis, it is necessary to discuss the benefits of this model for the investigation of dental inheritance. Potter and her co-workers stated most clearly their conception of the usefulness of path analytic models. To claim that phenotypic similarity between relatives is due only to genetic factors, wrote Potter *et al.* (1983), is to overlook any nonrandom familial environmental sources of variation, so that a positive parent–offspring or sib–sib correlation can only be viewed as evidence for a genetic mechanism. This *a priori* assumption, they noted, may lead to meaningless estimates of heritability, particularly in the evaluation of complex dentofacial developmental traits. Their incisive analysis led them to distinguish the following sources of familial covariance that may have direct or indirect influences on developmental variations of the odontometric phenotype:

1 Transmissible and nontransmissible environmental factors that are distributed nonrandomly among families or groups of families.
2 Correlations between genetic and environmental factors within individuals that may result in environmental covariance components between parents and their offspring.
3 Correlations between siblings who share a common maternal environment and the same postnatal family environment, where covariates such as socioeconomic status are the same.

Using the path analysis previously described, Potter *et al.* (1983) contrasted the MD dimensions of an early forming and stable tooth, the upper first incisor (UI1), to a late forming and variable tooth (UI2). Their results indicated that nontransmissible environmental factors were significantly relevant to the determination of the MD dimensions of both teeth. Noting that these dimensions were completed *in utero*, they concluded that prenatal rather than postnatal factors were involved. In their results, these authors also saw an explanation for previous findings of unequal environmental effect on tooth size between mono- and dizygotic twin types (Potter *et al.*, 1976), in that the intrauterine condition may have been a source of nontransmissible environment that was unique for each twin type as well as for singleton siblings.

Their paper also provided compelling information on the sex differences between the first molar (M1) and the second incisor (I2). Results for the former suggest that transmissible factors were important in their determination and also that there were no sex differences in the proportion of transmissible or nontransmissible environmental factors. Hence, it appeared that differences in mean MD dimensions between males and females could be ascribed to systematic cultural or biological differences. In contrast, results for the lateral incisor indicated that different nontransmissible factors were relevant to the development of this tooth in males and females. Hence no X-linked hypothesis was required for the explanation of first molar and lateral incisor size. In summary, it appears that molar size differences between males and females are due to systematic cultural and biological factors that do not contribute to within-sex variability, whilst sex differences in lateral incisor size are largely due to nontransmissible environmental factors.

### 3.8    Postscript

Careful attention to the distinction between environmental and genetic effects is crucial to a proper understanding of tooth size determination. The interplay between these two factors is subtle and interesting, and remains incompletely understood. On the one hand, there is a large body of dental genetic literature documented in the earlier sections of this chapter. On the other hand, there remains the nagging suspicion that nongenetic factors may have a significant effect in the determination of dental dimensions.

Of particular concern is the question of whether crown sizes may be altered by the maternal environment or by differential rates of foetal development. Consideration of prenatal factors that affect crown size has centred largely on chromosomal abnormalities (e.g. Cohen, Garn & Geciauskas, 1970, Garn, Cohen & Geciauskas, 1970), cleft palate syndromes (e.g. Garn, 1977; Dijkman, 1978) and on the maternal rubella syndrome (e.g. Evans, 1944, 1947; Bailit & Sung, 1968). Little attention has been paid to the effects that specific maternal and prenatal factors may have on dental dimensions.

In an attempt to elucidate these gestational variables, Garn, Osborne & McCabe (1979) examined 870 individuals of the 1963–66 birth cohort of participants in the Collaborative Perinatal Project. These children were studied at the time of their seven- to eight-year examination, along with a selected group of clinically normal subjects. The series included children of diabetic, hypothyroid and hypertensive mothers as well as low birthweight neonates and high birthweight neonates. From their survey it became evident that a variety of definable factors operating well before birth may affect crown dimensions. Accordingly, it was suggested that these findings reiterated the importance of the embryonic and foetal periods, even for structures such as the crowns of teeth that are completed postnatally. Specifically, it was found that maternal diabetes, hypothyroidism and large size at birth were associated with larger maxillary and mandibular teeth. On the other hand, crown diameters were diminished in maternal hypertension and in low birthweight conditions.

As Garn *et al.* (1979) noted, there are many disturbing implications to these findings. If small size at birth is associated with diminished crown size, then some part of the apparent human trend towards dental reduction may derive from the improved survival of small and developmentally immature neonates. Conversely, if large teeth are associated with large size at birth, then this may be a reflection of induced labour and deliberate attempts to minimise intrauterine foetal growth. Additionally, intra-individual crown size correlations, similarities between isomeres and antimeres and even odontometric fields might well derive more from specific intrauterine events at critical times during odontogenesis and less from genetic programming. Garn *et al.* (1979) further questioned whether conventional twin correlations were not artificially augmented by the inclusion of low weight, premature multiple births. Finally, population differences, geographical clines and secular trends may, to a greater or lesser extent, simply be a reflection of systematic prenatal influences on the widths and lengths of developing tooth crowns.

To repeat, the extent of the environmental effect on tooth size is ambiguous and remains difficult to assess. Family studies of crown size are at the mercy of sample size and low correlations. A further complication is introduced by missing teeth, which tend to be associated with diminished crown sizes of the other teeth (Garn, S. M., personal communication). What is claimed here, is not that the foregoing studies are all limited in value and significance, but rather that the field of dental genetics remains unexplored and ripe for exploitation.

# 4    Comparative odontometry

## 4.1    Introduction

It is probably a reflection of the limitations of the human brain that our thought processes are so heavily dependent upon the recognition of groups. To recognise something is to incorporate it instinctively into a category or group consisting of objects that share similar characteristics. An unclassified object tends to be mentally disturbing. Small wonder that the history of human odontometrics is so tightly interwoven with the much wider issues of the classification of living organisms.

From the earliest Babylonians there emerged a conviction that form, whether manifest in anatomy or music, was capable of numerical expression. This conviction was carried forward by the Pythagorean school of philosophers who, as early as the fifth century BC, sought to describe and compare animal forms numerically (Stapleton, 1958). Limited by their inadequate geometrical techniques, Pythagorean classifications soon lost ground to the dichotomous method of classification introduced by Aristotle in the fourth century BC, which emphasised the evaluation of form rather than measurement. Later adopted by Linnaeus (Cain, 1958), Aristotelian empirical syllogistic classifications maintained their ascendancy well into the present century when they coexisted, albeit often uncomfortably, with the more quantitative Pythagorean approach (exemplified for instance in the works of D'Arcy Wentworth Thompson).

The earliest efforts at odontometric classification were based on Aristotelian logic and consisted of attempts at discovering the essence (or 'real nature') of different ethnic groups. Aristotelian syllogistic logic favoured the consideration of properties that are essential or inevitable consequences of being; for example, the 'essence' of a tooth in a plane surface is best expressed by its definition as an object reduced to length and width. This type of reasoning, used by systematists of the stature of Cesalpino and Linnaeus, gave rise to the concept of 'analytical entities', such as species and genus, but unfortunately never quite lent itself to broader biological interpretation, which favours 'unanalytical entities' or properties that cannot be inferred merely from a few measurements (Sokal & Sneath, 1963).

The history of human odontometry may conveniently be divided into

30

three phases: after the initial descriptive phase there followed an essentially Aristotelian comparative or classificatory phase which, in turn, was succeeded by an interpretative phase. Typical of the descriptive phase is probably the first known human odontometric study, conducted in 1874 by Muhlreiter on the dentition of a skeletal population from Salzburg (see Lukacs, 1977). Revelling in the 'scientific objectivity' that the analysis of quantitative data conferred upon tooth measurements, late nineteenth and early twentieth century anthropologists relied heavily on odontometry to underscore the biological determinism which was so rampant at that time (Gould, 1981). As soon as comparative considerations were added to purely descriptive data, however, new questions arose. Could human groups be characterised by their teeth? Was tooth size an adequate and reliable discriminator of ethnicity? – and so on. An early, essentially Aristotelian, comparative evaluation of tooth size was that of Henry Fowler, who related the cranial base length to the overall length of the crowns of the maxillary cheek teeth into a dental index with which he divided human races into micro-, meso-, and megadonts (Fowler, 1885). In a similar vein, Maximillian de Terra (1905) suggested that the size of teeth was in proportion to the size of the body, and hence could be of no use in the distinguishing of human races. Arguing that families combine to form races, Miyabara (1916) was able to distinguish Japanese from European teeth by their crown sizes, thus setting a comparative odontometric trend that was to continue to the present day.

Comparative odontometry yields measures of affinity and dissimilarity in the same way as other metrical and nonmetrical traits do. It grew out of craniometry, from the pioneering work of Karl Pearson who introduced to physical anthropology the statistics of sampling and variance, the correlation coefficient and his Coefficient of Racial Likeness. Odontometric comparisons are, however, only as good as their underlying statistical techniques. Hampered by the limitations of univariate statistics and the tedium of their computation, early workers achieved little more than the evaluation of sample differences by using Student's *t*-tests. It is less than surprising therefore, that comparative studies had virtually ceased about a generation ago (Howells, 1984). As Howells emphasised, it was multivariate statistics that provided the escape.

Together with the publication of general multivariate texts such as those of Rao (1952), Anderson (1958) and Sokal & Sneath (1963), came the availability of electronic computers and their attendant packaged programmes. The new statistical technology brought its own misconceptions and problems, however. While there were those who revelled in the mathematical complexity of these techniques, accepting their results unconditionally, there were others who warned against the acceptance of the superiority of elaborate multivariate techniques over simpler quantitative distances (see Corruccini, 1973, 1975; Campbell, 1978). The hazards of over-use, or over-easy use, of canned statistical programmes, have thus to be weighed constantly against the benefits of the relatively greater robusticity of multivariate techniques.

## 4.2    Comparative odontometry

Comparative odontometric studies do not merely involve statistical carving up of samples into groups that are in fact nothing more than sections of a normal distribution (Howells, 1984). Central to the subject of comparative odontometry, lie the dual concepts of discrimination and allocation. Discrimination involves the study of ways in which groups differ with respect to one or more characteristics or traits. Not only is the researcher engaged in seeking for ways to 'discriminate' between the groups, but he also wishes to know how well they discriminate, or which characters are the most powerful discriminators. The characters used to distinguish among groups are called discriminating variables. Because they have to be measured at the interval or ratio level, odontometric data lend themselves particularly well to this type of investigation.

The problem of allocation, on the other hand, consists of classifying an unknown individual into one of a number of populations on the basis of measurements derived from a given set of characteristics. The following example will illustrate the nature of this problem. A forensic scientist is presented with a human mandible and all its teeth for identification. Assuming that somehow the ethnicity of the remains is known, he now wishes to sex the individual. Although the problem now reduces to the allocation of an individual into one of two populations (male or female), he also needs to estimate the probability of misclassification. This is particularly important if there is an area of overlap between the multivariate distributions of the two sexes. Prior knowledge of which diameters are the best discriminators would also be useful to him, especially if he knew that he could rely on one or two measurements only.

The process of allocation becomes a very difficult one when an individual has to be assigned to any one of a number of ethnic groups of either sex. The present chapter examines some of the uni- and multivariate methods that have been employed for discrimination or allocation based on odontometric data. Minimal attention will be paid to the mathematical complexities surrounding these techniques, as these are adequately covered by standard statistical texts.

**4.3    Odontometric discrimination I: population, means and the *t*-test**
Statistical discrimination deals with differences between populations and samples drawn from populations. Statisticians refer to a population as any well defined collection or set of values and a sample as any subset of values that has been drawn from a population. This sense of the term population thus differs from the sense in which most scientists use the term. Typically, the scientist's population consists of a collection of specimens, people, laboratory animals and so on, and may give rise to many populations in the statistician's sense of the word.

The manner in which samples are selected from a population is critical to the process of statistical discrimination, since sampling procedure determines the way in which chance factors affect the statistics with which we are concerned. There are two broad classes of sampling processes – probability sampling and random sampling. In the former, chance factors determine which elements in the population will be included in the sample. Random samples result when each element in the population has an equal probability of being selected and when the choice of an element has no effect on the choice of any other element. In general, when a statement is made that a statistical method is based on the assumption that the sample is random, what is meant is that the sample meets the requirements of independence and equal probability. It is desirable that distance analyses based on odontometric data should be calculated from random samples.

As we have seen in the previous chapter, the distribution of a sample may be represented by the classic bell-shaped curve depicted in Figure 3.1. It is worth noting the following characteristics of such a normal distribution:

1 It may be specified by two parameters: its mean and its standard deviation (or its variance).
2 It has a continuous and symmetric distribution.
3 The curve is asymptotic, which means that the graph extends out into infinity without ever touching the horizontal axis.

Suppose there are two populations, each normally distributed, with means $\bar{x}_1$ and $\bar{x}_2$ and variances $\sigma_2^1$ and $\sigma_2^2$ respectively and a sample is drawn from each. The difference between these two population means, $\bar{x}_1 - \bar{x}_2$, is an example of a population comparison. The question is whether there is a significant difference between the mean values of the two populations from which the samples were drawn. Hence we need to distinguish between two possible situations: first, the two means are so close to one another that the difference between them may be considered to be due to chance fluctuations in sampling (sampling error), or secondly, the difference between them is so great that it can no longer be attributed to sampling error. How probable is it that our two samples represent two identical populations? A hypothesis, called a null hypothesis, is now set up to compare the two means. This hypothesis states that the two samples come from two populations whose means $\bar{x}_1$ and $\bar{x}_2$ are equal. A statistical test is required which will show with what confidence the null hypothesis can be accepted or rejected. One of the most sensitive, useful and powerful tests of the null hypothesis was formulated by William Gosset in 1908, under the pseudonym Student: hence it is called Student's $t$-test. Formulae for its calculation are available in any standard statistical handbook. Suffice to say that the test is based on the fact that if many pairs of random samples are drawn from the same population, their mean values will differ only by chance. Thus the mean of all the differences between the means is zero, and the distribution of the mean differences will be bell-shaped (or Gaussian). However, if the difference between the two samples is much larger than zero, it may be possible to reject the null hypothesis that the difference between the means is zero. The $t$-test, defined as the ratio of the difference between the means, divided by the standard error of the difference, will provide a useful method of determining the distribution of the mean differences.

Much information on population differences in tooth size has been gleaned from straightforward comparisons using $t$-tests. For instance, in their classic early study, Barrett, Brown & MacDonald (1963) used $t$-tests to show that Australian Aboriginals had significantly larger teeth than ten other populations. Similarly, van Reenen (1966) compared the mean dental diameters of nomadic Bushmen with those of sedentary Bushmen and found that the latter had larger teeth than the former. Ghose & Bahdady (1979) used $t$-tests to compare metric data of three peoples living in the Arabian Peninsula. Their results indicated a size gradient from north to south, with Iraqi teeth being demonstrably larger than those of Bedouins or Yemenites. Axellson & Kirveskari (1983) extended the use of $t$-tests in odontometric analyses by comparing Icelandic tooth size with that of other populations. The crown sizes of Icelanders proved to be larger than those of European and North American caucasoids, being exceeded only by those of Australian Aboriginals.

The major emphasis of a similar study by Lukacs (1985) centred on odontometric comparisons between prehistoric Indian populations. Tooth crown measurements for a late Chalcolithic skeletal series from Inamgoan were contrasted to data from other prehistoric populations in India and Pakistan. The Inamgoan were found to have larger teeth than the others, which was interpreted as being a biological adaptation to coarse dietary items and primitive food preparation techniques. Recently, Calcagno (1986) contrasted tooth size changes among Nubian archaeological populations dating from the Mesolithic through the Christian era.

Student's *t*-tests revealed significant reduction in dental size between Mesolithic and more recent samples in Nubia (to be discussed later). Like those before him, Radnzic (1987) turned to *t*-tests when he investigated interpopulation differences in tooth size. His comparison of indigenous British and immigrant Pakistani teeth showed that there were no significant differences between the two ethnic groups.

## 4.4     Odontometric discrimination II: size and shape

The previous section emphasised the usefulness of the *t*-test as an indicator of the statistical significance of differences between two means. However, it has its limitations. Suppose one wished to evaluate population differences in terms of a number of characters, for example, to compare indigenous British and immigrant Pakistani teeth, not on a tooth by tooth basis, but rather on a mouth by mouth basis (i.e. 32 variables, each with its own mean value). One of the simplest ways of doing this would be to calculate a 'distance' between the two populations. Because we intuitively think in terms of Euclidean geometry, the most familiar measure would be a simple Euclidean distance. In practice, however, Euclidean distances have been shown to be poor estimators of interpopulation dissimilarity (Dunn & Everitt, 1982).

In 1954, Penrose suggested dividing the Euclidean distance into two parts, a coefficient of 'size' and a coefficient of 'shape'. Penrose defined the size of a trait as the mean of its character values and the 'size' distance between the two forms as the square of the difference in their sizes. The 'shape' coefficient was taken as the variance of differences between the character values of the populations being compared.

Penrose's shape and size differences found their first use on human odontometric data in the work of Doran & Freedman (1974), who reported odontometric differences among a number of populations from Oceania and Australia. Using Penrose size and shape metrics, these authors were able to show that Goroka and Lufa populations of the eastern New Guinea highlands had megadont dentitions similar to those of Aboriginals, but smaller than the Nasioi of Bougainville Island. A more general point about the usefulness of Penrose distances in odontometric studies was made by Harris & Nweeia (1980b). They reported that when Penrose metrics were used to evaluate phenetic relationships between various Amerindian groups, shape rather than size emerged as the better discriminator. Several authors have in fact commented on the potential benefits of basing systematic analyses on shape rather than size. Corruccini (1973) argued that distances between populations should not be seen purely as measures of differences in size. Since shape may be defined as differential size and since differences in size will tend to outweigh differences in proportion with

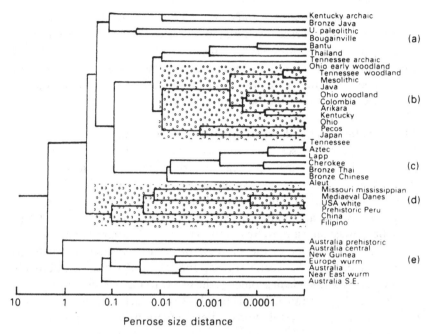

Figure 4.1. A dendrogram based on Penrose size distances computed from human odontometric data. (a) Upper Paleolithic and megadontic; (b) Mixed economy; (c) Technologically advanced; (d) Agriculturally based; (e) Least technologically advanced. (Adapted from Perzigian, 1984.)

increasing taxonomic distance, he argued that shape will be the more important determinant of dissimilarity in closely related populations. Support for this view has since come from Corruccini (1975, 1987), Campbell (1978), Releford (1984) and Thorpe & Leamy (1984).

Separate consideration of size and shape was given by Perzigian (1984) in his evaluation of human odontometric variation. A dendrogram based on Penrose size distances indicated to him that spatially, temporally and genetically varied groups of people may be clustered together on the basis of odontometric data (Figure 4.1). As this figure illustrates, the populations investigated fell into five clearly defined clusters which Perzigian felt reflected their technocultural advancement. This suggested to him that there may well be an association between tooth size and technocultural factors. When he considered Penrose shape coefficients, however, a different picture emerged. Shape distances tended to follow phenotypical rather than cultural classifications.

The use of Penrose coefficients in human odontometric distance analysis has recently been investigated by Groeneveld & Kieser (1987). Using a computer intensive technique (Efron, 1979) these authors evaluated the statistical accuracy of a battery of different metrics. Particular attention was paid to stability in the face of reduced sample size and to statistical bias resultant from over- or underestimation. Their results clearly indicated that Penrose's shape coefficient was the most reliable metric examined. The usefulness of Penrose's coefficients remains hampered by the fact that statistical evaluation of differences between individual Penrose distances remains impossible without confidence intervals for these values. This problem has subsequently been addressed by Groeneveld & Kieser (1988) who described a modified procedure by means of which standard errors may be generated for Penrose values.

Current debate centres not around the value of shape measures such as those of Penrose, but on two other problems: first the definitions of size and shape and, secondly, their relative importance. While it is clear that size may easily be defined as the magnitude of a vector of measurements, the definition of shape remains elusive (see Corruccini, 1987). Definitions of shape range from the almost sublime (... a ray in multidimensional hyperspace – Mosiman & James, 1979) to the esoteric (... an outline of form from which scale information has been drained – Bookstein, 1978). As representative of the second controversy, one might cite Atchley (1983) who asserted that previous morphometric investigations had assumed size to be a random noise component which had to be removed from their analyses. Corruccini (1987) disagreed, suggesting that size information needs to be separated but not discarded from shape. For these and other reasons, it is probably necessary at the present time to continue to distinguish between size and shape in dentomorphological analyses.

## 4.5    Odontometric discrimination III: multivariate techniques

In contrast with the more classical approach of comparing mean values of one variable in two populations (e.g. the Student's *t*-test), multivariate statistical analysis allows for the study of several variables at once. This, in turn, provides the investigator with a greater efficiency and a greater power of interpretation, especially in the analysis of large sets of data. When speaking of multivariate analysis, it is important to have in mind the use of several variables, whose simultaneous correlation among themselves is as important as their collective difference from other sets of measurements.

The use of pairs of measurements to locate points in a single plane or triplets of numbers to locate points in three-dimensional space was first spelt out in the mid-seventeenth century. Towards the latter half of the nineteenth century, mathematicians began to realise that there was no need to stop with triplets of numbers. It was recognised that quadruples of numbers $(p_1, p_2, p_3, p_4)$ could be regarded as points in four-dimensional space. Similarly, quintuples $(p_1, p_2, \ldots p_5)$ could be viewed as points in five-dimensional space. Of course, human powers of geometric visualisation do not extend beyond three-dimensional space. It is, nevertheless, possible to extend many concepts beyond three-dimensional space by working with the numerical properties of points and vectors rather than with the more familiar geometric properties. Thus if $p$ measurements have been made on $n$ individuals, we may represent each individual as a point defined by these measurements, and lying in a $p$-dimensional Cartesian space. Multivariate techniques basically provide methods of changing our vantage point of looking at these $p$-dimensional data spaces. We may, for instance, wish to visualise the dispersion of data points within each cloud of points (e.g. principal components analysis), or we may wish to determine the best vantage point from which to view differences between two or more sets of clouds (e.g. canonical analysis).

Consider the case where MD and BL measurements are obtained on the first right upper molars of ten individuals. Each individual may then be represented by one of ten points in a two-dimensional space consisting of two axes at right angles to one another (i.e. orthogonal). The ten individuals will be represented by a cloud of points on the bivariate graph (Figure 4.2). If the two measurements are correlated, the cloud will be elliptical or cigar-shaped. The ellipse will, of course, represent a size gradient from individuals with small teeth in the lower left to those with large diameters in the upper

Figure 4.2. A bivariate plot of ten hypothetical MD and BL measurements. The elliptical cloud of data points indicates a correlation between the two variables, with small toothed individuals occupying the lower left and large toothed individuals the upper right.

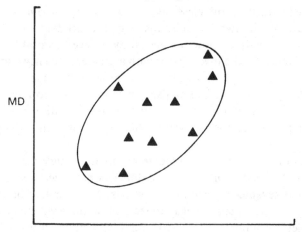

MD

BL

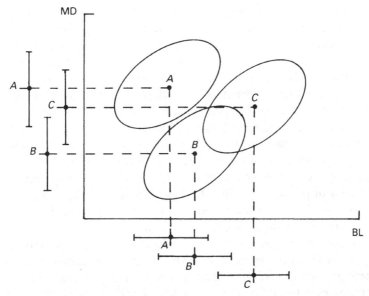

Figure 4.3. A bivariate plot of the hypothetical MD and BL dimensions of first upper molars in three different populations. Approximate 90% confidence ellipses denote intragroup dispersions. Means and standard deviations are given next to the respective axes. (Adapted from Albrecht, 1980).

right. Although individuals who share a common or similar morphology will occupy similar positions within the two-dimensional space, examination of relative positions may provide misleading information as to the separation of morphologically similar individuals. This is best illustrated by the graphical representation (Figure 4.3) of hypothetical data on first upper molar dimensions. Here, the bivariate plots of each of three populations appear as three elliptical clouds of points, with dimensionally similar individuals sharing similar positions within the data space. Considering the MD dimensions only, populations A and C emerge as the most closely related groups. Consideration of the BL dimension, however, leads to a different biological interpretation: populations A and B are the most closely related. Moreover, the three ellipses are orientated such that B and C share an overlapping morphology in that large toothed individuals of population B have similar teeth to small toothed individuals in population C. Population A in contrast, is easily distinguished by its different relationship of the two measurements.

The foregoing discussion illustrates two very important concepts. First, bivariate data may be used to achieve an objective representation of the morphological relationships between the three populations. Secondly, the determination of the biological significance of these relationships is subjective, essentially independent of the morphology itself (Albrecht, 1980). In order to make sense out of multiple measurements used to describe more than one population group, one has to turn to other methods which allow for the simultaneous consideration of the joint distribution of these variables. These techniques allow data to be examined in space of several dimensions. For instance, instead of looking at the three populations in the two-dimensional space of Figure 4.3, one might wish to view the relations among these three clouds of points from another vantage point. The change of vantage point to a different set of axes may be achieved by the rotation of the system of clouds. Rotation that emphasises the overall dispersion of points, for instance, is used in principal components analysis. On the other hand, rotation that emphasises the distances between individual clouds is used in canonical analysis.

Both these techniques are useful in the analysis of odontometric data and computer software is commonly available for their calculation. Discussion of these methods, however, will be limited to a brief exposition of their underlying principles. For the reader who seeks a mathematically more complete presentation of the statistics involved, there are several useful textbooks – for example Rao (1952), Sokal & Sneath (1963), Cooley & Lohnes (1971), Overall & Klett (1972), Klecka (1980) and Reyment, Blackith & Cambell (1984).

### 4.5.1    *Principal components analysis*

Considering two data points within one of the populations plotted in Figure 4.3, the distance between these two points (say $A$ and $B$) may of course be represented by a straight line. If we now reduce the number of dimensions by projecting these two points onto the abscissa, for example, then the actual distances between the two points will be grossly misrepresented (Figure 4.4(a); $a_1 b_1$). However, a projection onto a line such as $\Delta$, which has its orientation in the same general direction as the overall scatter of points, will preserve the distances between points much better (Figure 4.4(b)). Unfortunately, if we replace point $A$ by its projection ($a_2$ onto line $\Delta$, we lose the information about distance $Aa_2$ (from $A$ to $\Delta$). From a practical point of view, we wish to choose the line $\Delta$ in such a way as to lose as little information as possible. The line that results in the least loss of information is taken to be the line for which the sum of the squares of each of the distances is a minimum. This is called the least squares regression line.

If we now wish to transform a set of points in multidimensional space into a set of points in one-dimensional space, we calculate a straight line, $\Delta$, such that the sum of the squared distances between it and the data points is a

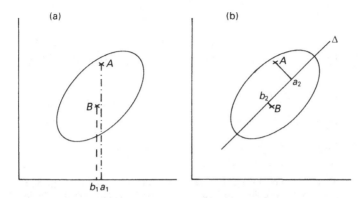

Figure 4.4. A reduction of dimensions by means of projection. (a) two points (*A* and *B*) projected onto the abscissa results in a gross misrepresentation of the actual distance between them. (b) projection onto a line Δ will preserve the distances between points much better.

minimum. This line may then be referred to as the first principal component (or axis) of the swarm of points. Basically, we have transformed space of any dimensionality in such a way as to lose the minimum information. If the swarm of data points is visualised as occupying a cigar-shaped cloud, then the first principal component gives an idea of where the long axis of the cigar lies. In order to visualise the thickness of the cigar, it will be necessary to project a line along which the entities have the second most spread, at right angles to the first axis. This line is called the second principal component (or axis). Because the values of the data points as measured in terms of these two axes are independent of one another, the axes are described as being orthogonal. Principal components analysis thus involves rotation of the original axes in such a way as to maximise the variation of the original observations. It is a valuable statistical tool, especially when the first principal component explains much of the variation of the scatter plot.

While principal components analysis enables the relationships between sets of measurements to be expressed in a space of fewer dimensions than that into which the measurements were originally plotted, care has to be taken in the interpretation of these results. Consider, for instance, the three populations depicted in Figure. 4.5. With respect to the first principal axis, the existence of three separate populations would not be suspected, mainly because their projection onto the first axis would produce a continuous frequency distribution. Projection onto the second principal axis, however, reveals three groups of individuals, each projected onto a different section of the axis.

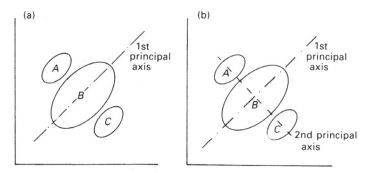

Figure 4.5. Possible misinterpretation of the results of principal components analysis. (a) The first principal axis produces a continuous distribution, masking the existence of three separate populations. (b) projection onto the second principal axis, however, reveals their identity.

### 4.5.2    Canonical analysis

Principal components analysis involves rotation of the original coordinate axes, such that the data space is viewed from a new vantage point which emphasises within-group variability along its major axis (Figure 4.6(a) and (b)). The data space may also be viewed from a vantage point which emphasises between-group variability. One such method is called canonical variate analysis, and it involves the determination of linear combinations of the original variables in such a way that the differences between reference groups are maximised, relative to variation within the groups (Campbell & Atchley, 1981). Canonical variate analysis quite simply involves a series of rotational and rescaling transformations of the original variables which, whilst preserving the integrity of the original data, makes them more amenable to interpretation. The complex original data space is transformed into relative simplicity by partitioning the total variability into within-group variability and between-group variability. The first step is to rotate the original coordinate axes so that they come to lie parallel to the major and minor axes of the within-group dispersion ellipses (hence parallel to the first and second principal components). Next, the variates are so rescaled so as to standardise within-group dispersion along all axes within the transformed data space (Figure 4.6(c)). The third step involves rotation of the now rescaled, standardised axes, so as to orientate them parallel to the major and minor axes of the between-group variation (Figure 4.6(d)). As Albrecht (1980) emphasised, canonical analysis does not involve extraordinary mathematical interventions in its computation.

Potter (1972) discussed the different results obtained when using univariate and multivariate techniques on human odontometric data. Stressing the importance of using the latter in the analysis of multiple and correlated variables such as tooth size measurements, she showed that the differences between sample means obtained by multivariate methods often differed from those obtained by univariate methods. Subsequently Potter and her associates (1981) investigated male–female separation by uni- and multivariate means in a sample of Tagalong Filipinos. From their univariate analysis, considerable sex dimorphism emerged for most of the variables studied. However, when correlations among the teeth were taken into account through multivariate analysis, male–female distance was found to be small, with substantial overlap between populations. These results prompted a call for a combined uni- and multivariate approach in the evaluation of odontometric data, which has since been answered by Kieser and his associates (1985d, e). In their metrical analysis of Lengua Indian teeth, these authors reported significant sex differences in 23 of the 26 variables considered, when using Student's *t*-tests. However, a number of these variables were found not to contribute significantly to sex differences in multivariate analyses. Similar results in an analysis of metrical data derived from whites underlined the importance of Potter *et al.*'s. (1981) combined approach to odontometric data (see Chapter 6).

Falk & Corruccini (1982) used uni- and multivariate analyses to evaluate the efficacy of traditional and non-traditional dental measurements in the sorting of human populations. It emerged that univariate analysis pointed to nontraditional, cervical, measurements as being the more reliable.

Figure 4.6. A sequence of procedures for the transformation of raw data into canonical variates – see text for discussion. (Adapted from Albrecht, 1980.)

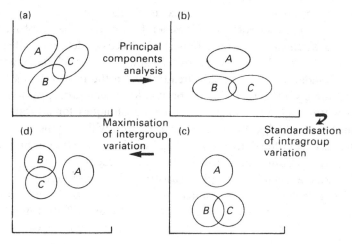

Multivariate techniques, on the other hand, showed that traditional maximum diameters were better at sorting human populations. Though these findings are difficult to explain, this study again demonstrates the importance of using both analytical techniques when analysing human odontometric data.

The usefulness of multivariate techniques in the analysis of interpopulation differences has recently been emphasised by Bhasin *et al.* (1985). These authors investigated odontometric data from various regions of India and found that multivariate analyses allowed for the separation of agriculturist groups from more tribal Indians from the west of the country and also from the highest caste Brahmans.

Multivariate methods have also been employed in the analysis of environmental sources of variation in human crown size (Mizoguchi, 1980), in the interpretation of the influences of earlier developing teeth upon later developing teeth (Mizoguchi, 1983a), and in the evaluation of age related tooth size variation in the Lengua Indians of Paraguay (Kieser *et al.*, 1985a). These results will be discussed in later chapters.

### 4.6     Odontometric allocation

In the previous sections, the description of group differences based on odontometric data have been dealt with. Discrimination is, however, only one side of the coin. Whilst the comparison of population differences will remain of importance to some physical anthropologists, it is probably fair to predict a lessening of interest in such pursuits, directly proportional to a growing awareness of the artificiality of classification of human populations. The other side of the coin presents a more challenging and an increasingly important facet of odontometric analysis. With the demise of discrimination, there has come a renewal of interest in allocation. The allocation of an unknown individual is a relatively unexplored field of immense practical importance. Consider, for instance, the problems facing a forensic scientist or an archaeologist who needs to determine the affinity of newly uncovered human remains. The ability to sex such remains, within given confidence limits, would already lighten the investigator's burden. Unfortunately, little attention has been paid to such problems, with many investigators actually using discriminant techniques for allocation procedures (Campbell, 1984). The present section will examine some of the problems encountered in allocation, and will introduce briefly a number of statistical procedures designed to overcome these obstacles.

Consider a bivariate sample consisting of males (population 1) and females (population 2), represented by centroids *A* and *B*, and associated with the 95% probability ellipses given in Figure 4.7. In this idealised representation it is assumed that the male and female covariances are

Population 1

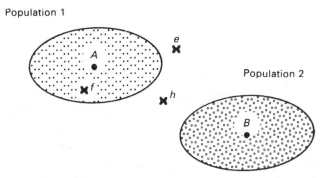

Figure 4.7. Idealised representation of a bivariate sample consisting of males (population 1) and females (population 2), represented by centroids $A$ and $B$ respectively and associated with the 95% probability ellipses illustrated. Individuals $e, f$ and $h$ are to be assigned.

known and equal and, also, that a new individual to be assigned has an equal *a priori* likelihood of belonging to either sex. Let us now consider three individuals ($e, f$ and $h$) to be assigned to either population 1 or 2. Knowing that each of these assignees has to belong to one of the populations, one could simply place each individual with that population to which it was the closer. Because a definite allocation decision has to be made, this process is referred to as a forced allocation procedure (Campbell, 1984). In this example, individuals $e$ and $f$ would clearly belong to sex 1, since they are closer to centroid $A$ (as measured, for instance, by means of Mahalanobis' distance). The situation concerning individual $h$, however, is less clear-cut. Although forced allocation would assign this individual to population 1, to whose centroid it is marginally closer, it should really be classified into an area of doubt.

Another method of allocation involves the linear combination of the original variables in such a way as to maximise the distance between the centroids of each sex, relative to intrasex variation. Here, the population centroids ($A$ and $B$) are projected onto a discriminant vector – $\mathbf{c}$ (Figure 4.8). In this case, points $\bar{Y}_a$ and $\bar{Y}_b$ represent the discriminant means for the two sexes. When a new individual is assigned to either sex, a discriminant score is determined in the same way, and this is then compared to the discriminant means. In the case of a forced allocation, the new individual is assigned to the population to which its discriminant score is the nearer. Under the assumption of equal *a priori* probability, point $s$ (Figure 4.8), represents the cut-off boundary between the two sexes. New individuals $e$ and $f$ are closer to population 1, because their discriminant scores ($q$ and $r$) are nearer $\bar{Y}_a$. The discriminant score for individual $h$ is also nearer $\bar{Y}_a$ and hence assignment will be with sex 1.

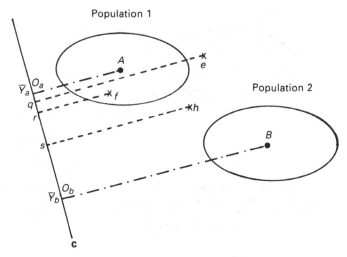

Figure 4.8. A representative of the discriminant function for the two populations given in Figure 4.7, with the vector c as the discriminant vector. (Adapted from Campbell, 1984.)

Canonical variate analysis provides another method of assignment. This procedure also involves the determination of linear combinations of the original variables in such a way as to maximise the differences between the populations, relative to within-group variation. The variates are, however, rescaled to standardise within-group dispersion along all axes within the data space. These axes are then rotated until they are parallel to the major and minor axes of the between-group variation (Albrecht, 1980; Campbell & Atchley, 1981). Figure 4.9 illustrates a linear discriminant function obtained for the two sexes by means of canonical analysis. Note that the distance between the canonical centroids ($C_1$ and $C_2$) is equal to the interpopulation Mahalanobis distance, and that the 95% concentration ellipses have been transformed into circles. The discriminant function now passes through the respective population canonical centroids (Campbell, 1984). Unfortunately, although individuals $e$ and $f$ are still closer to canonical centroid $C_1$, individual $h$ remains of doubtful allocation, nearly halfway between $C_1$ and $C_2$. Thus, in order to make the allocation of new individuals more meaningful, further assessments need to be undertaken.

The first of these concerns the relative affinities of a new individual for each of the reference populations. Such a probability of allocation is referred to as a posterior probability of membership and is related to the probability density function of the given vector, for each of the populations (Campbell, 1984). Posterior probabilities may be determined according to the method described by Campbell (1984) and by Kieser & Groeneveld

(1988e). Consider the allocation of a new individual *e* in Figure 4.9. For this assignee, the squared Mahalanobis distance may be partitioned into two components. The first is due to its distance in the direction of the discriminant vector (i.e. $C_1 - q$), whilst the second is due to its distance orthogonal to the discriminant vector (i.e. $q - e$). Thus its distance from population 1 may be written as:

$$\Delta_{1,e}^2 = (C_1 - q)^2 + (q - e)^2$$

and the distance from population 2 as:

$$\Delta_{2,e}^2 = (C_2 - q)^2 + (q - e)^2,$$

with $(q - e)$ common to both populations. Hence it is the first component which is relevant to the calculation of the posterior probability. This may be written as $C_k - q$, which is the deviation of the discriminant score for the new individual from the canonical centroid of the *k*th population. The posterior probability determines the affinity of the assignee to a reference population, with consideration of the other populations.

Alternatively, one might consider the affinity of a new individual for each population in turn, without consideration for the other populations. Such probabilities are referred to as typicality probabilities of membership. Campbell (1984) has suggested that a typicality probability for the *k*th

Figure 4.9. A linear discriminant function obtained for the two sexes by means of canonical analysis, with canonical centroids $C_1$ and $C_2$ for males and females. Axes PCI and PCII represent orthonormal linear combinations of the original variables. (Adapted from Campbell, 1984.)

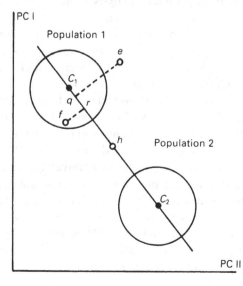

population may be seen as the probability of finding data vectors which have probability density values greater than those of the given data vector for the individual to be assigned. Thus, according to Campbell (1984), for the multivariate normal distributions assumed here, the typicality probability may be calculated by referring an individual Mahalanobis-$D^2$ distance for the $k$th population to the chi-squared distribution, on $v$ degrees of freedom ($v$ being the number of variables observed for each individual).

In Figure 4.9, individual $f$ lies within the 95% concentration contour, and hence would be considered typical of population 1. However, individual $e$ lies outside the 95% concentration contour for this population and would thus be associated with a nonsignificant typicality. The Mahalanobis distance for this individual, as we have seen, is given as $C_1 - e$. There are two components to this distance, one in the direction of the discriminant vector ($C_1 - q$) and another in the direction orthogonal to the discriminant vector ($q - e$). In the case of individual $e$, the latter component is larger than the former and hence it has a small typicality probability.

## 4.7    Allocation based on human odontometric data

Consideration of human odontometric data has centred largely on univariate and multivariate discrimination between sexes and between ethnogeographic groups. In contrast, little or no attention has been paid to the allocation or assignment of individuals on the basis of dental data. Campbell (1984) and, more recently, Kieser & Groeneveld (1988e), stressed the importance of using appropriate statistical methods for these two independent, yet often confused, procedures. Of particular concern has been the inappropriate use of discriminant analysis in the determination of group membership of individuals. Yet the problems that beset allocatory procedures are central to the fields of palaeo- and physical anthropology, and to forensic odontology. For this reason, Kieser & Groeneveld (1988f) decided to conduct an empirical evaluation of allocation and discrimination based on human odontometric data.

These authors contrasted the confidence with which individuals could be grouped and then allocated on the basis of tooth size. Their data were derived from MD and BL diameters of 202 Lengua Indians (100 male, 102 female), 125 caucasoids (59 male, 66 female) and 206 Negroes (106 male, 100 female). Basically, the aim was first to effect multivariate intergroup discrimination by means of canonical and stepwise discriminant analysis; secondly, to evaluate the confidence with which individuals could be reallocated into their original groups. Their results showed that male and female Negroes, caucasoids and Amerindians could be reliably separated by multivariate means. In fact, even after reduction of bias by means of a jack-knifing procedure, correct classification could be anticipated in 66% to

78% of cases. In sharp contrast, the results indicated that a low proportion (between 12.5% and 55.4%) of individuals could be allocated with a high degree of confidence, even though it is known *a priori* that they were drawn from one of the original populations. Further, a high proportion (between 14.3% and 43.5%) of individuals were highly atypical of the very populations from which they were drawn. These results, it was suggested, cast a very gloomy light on the usefulness of odontometric data in the allocation of fossil or forensic specimens. However, they do provide a further illustration of the very considerable variation that exists among individuals within a population. Whilst multivariate discriminant procedures may be usefully employed in the separation of different populations, allocatory procedures indicate that individuals cannot be assigned with the same degree of confidence, even when their *a priori* membership is known. In addition to these implications, the results also relate to the more general issue of the use of discriminant techniques in the allocation of individuals. These two processes are exclusive, each one requiring a set of manipulations which are uniquely suited to it alone.

It may abstemiously be concluded from the foregoing, that the subject of allocation is urgently awaiting contemplation. The results of such researches will benefit scientists of many different persuasions. Not only will the refinement of allocatory procedures aid in solving some of the questions of comparative anatomy, but it will serve as a test-bed for many existing craniofacial evolutionary hypotheses.

# 5 Odontometric trends

## 5.1 Introduction

The study of human odontometric trends relies upon repeated measurements of tooth size in a given population. Whilst short term trends may be evidenced by measurements taken on samples a human generation apart, the identification of long term trends will involve samples separated by thousands, or even millions of years.

Repeated dental measurements may show the tooth size of a population to be either static or varied. For teeth to remain the same size over a long period of time, two explanations may be proposed. One is that tooth size cannot readily be changed by selection because of restrictions imposed during ontogeny (developmental constraints), and the other is that individuals who have larger or smaller teeth are of lower fitness (stabilising selection). Tooth size variation over a long period of time may also be explained in terms of natural selection.

Given that natural selection will tend to preserve the most favourable variants within a population, it could be concluded that a change in the environment will pressurise a population towards the achievement of greater adaptation. Hence, if a change in tooth size enables a population to improve its ability to meet the challenge of a new environment, this change will be favoured by selection, and one may speak of a phyletic or evolutionary change in tooth size.

Slow but persistent alteration over a shorter period of time is referred to as a secular change. In human biological terms, the secular trend relates to those systematic short-term changes that occur in successive generations of adults, adolescents, children and neonates in a continuing population. Although traditional investigations have focused on problems relating to stature, weight and those events that surround the onset of puberty, growing numbers of other metric traits are being investigated (Tobias, 1978, 1985). A positive secular trend is one that demonstrates an increase in the dimensions under consideration. In contrast, a negative trend relates to a diminution of the structure under consideration. It is therefore of the utmost importance that the direction of secular processes should be specified (Tobias, 1985).

**5.2     Negative odontometric trends**
Change in size is probably the most frequently documented evolutionary phenomenon, size increase being the most prevalent trend. However, bigger not necessarily being better, large structures have on occasion developed into pygmy forms. Because the dimensions of homologous structures such as teeth can be measured with such a high degree of precision, these dimensions have provided scientists with abundant and reliable data on which to base their quantifications of phyletic change. The most pervasive and also the best documented trend in human tooth size since the Palaeolithic has been one of reduction (Dahlberg, 1963; Brace, 1967, 1976, 1979, 1980; Wolpoff, 1969; Brace & Mahler, 1971; Brose & Wolpoff, 1971; Le Blanc & Black, 1974; Frayer, 1978; Chamla, 1980; Brace & Nagai, 1976; Lukacs, Joshi & Makhija, 1983; Kieser *et al.*, 1985a; Lukacs, 1985; Calcagno, 1986; Macchiarelli & Bondioli, 1986).
There are two general theories as to its cause. The traditional one explains tooth size reduction in terms of active selection for smaller teeth, whilst the second principal explanation suggests that tooth size reduction is the result of reduced selection.

*5.2.1     Selection for smaller teeth*
Selective models of dental reduction have sought to interpret smaller tooth size as an advantageous condition differentially preserved by natural selection (Lukacs, 1985). Three major factors relating to the adaptedness of smaller tooth size are (i) the energy budget effect, (ii) the reduction of jaw size and (iii) body size reduction.

*Energetic budget effect*
The energetic principle of evolution dictates that evolutionary restructuring will be ineluctably linked to the balance of energetic costs. Thus whilst structural reduction will be accompanied by a decrease in energy expenditure, an increase in size will inevitably be associated with increased energy demands (Peters, 1983). This means, in effect, that size reduction may be interpreted as being an adaptation with respect to the energy budget of the organism. Dullemeijer (1980), however, stressed that the energy budget principle remains a theoretical model in evolutionary arguments, largely because of the difficulty of visualising the entire energy balance of the organism. Moreover, evolution should not be seen merely as a strategy for conserving energy. Rather, phylogenetic changes in tooth size should be seen in the light of the organism's capability to gather food and to convert it into energy. Improvements in the gathering mechanism will

include modifications that increase the usefulness of teeth as a means of procuring or processing food.

In early human terms this would mean an increase in the efficiency of teeth as tearing, holding and breaking instruments. Such modifications would result in the enlargement of dental material. Improvements in the converting mechanism, on the other hand, will focus on greater digestive efficiency per unit of tooth substance and would thus result in tooth size reduction. In any given group of individuals, the simultaneous optimisation of these two processes will result in a delicate balance between dental enlargement (energy expensive) and dental reduction (decreased energy cost). It could be argued, therefore, that with the selective balance in post-Pleistocene humans will have swung in favour of tooth size reduction. Individuals with smaller teeth will have gained greater reproductive success, not because they produced more offspring, but because their children were produced at a lesser energetic cost.

### Phylogenetic facial shortening

That phylogenetic shortening of the maxillofacial complex may have led to a reduction in tooth size was first proposed by Charles Darwin (1872) and subsequently by Hellman (1936), Robinson (1954) and Buettner-Janusch (1966). Riesenfeld (1970) showed experimentally that shortening of the face produces crown size reduction in laboratory rats. Dental reduction coupled to facial shortening does imply a correlated response to natural selection between teeth and the jaws that house them. Hence it has to be accepted that selection for a particular trait (e.g. shorter jaws) will have an indirect effect on the distribution of correlated characters such as the sizes of individual teeth.

The measurement of selection on correlated characters was extensively dealt with by Lande & Arnold (1983). These authors showed that it is possible to distinguish between directly selected characters and indirectly selected traits, which change only owing to their correlations with directly selected characters.

They cautioned, however, that such analyses were valid only if the characters under observation were phenotypically uncorrelated to other traits that might be subjected to directional selection. Hence all inferences about correlated responses to selection in the orofacial region must be drawn from comparisons of phenotypes within one population, over a long period of time.

### Body size reduction

Tooth size reduction could be viewed as a correlated response to overall body size reduction. One such theory is that of Macchiarelli &

Bondioli (1986), who focused their attention on dental changes in Late Pleistocene humans. These authors noted that the gradual reduction in human dental structure showed a marked acceleration during the transition from Pleistocene to Holocene, after which the trend diminished. During this period similar changes in stature were noted. They reasoned, therefore, that the transition from Late Pleistocene hunting and gathering conditions to those involving larger, more sedentary groups had introduced stresses that were important in producing transitory reduced expressions of the genetic background. This, they postulated, could be ascribed to the two factors acting simultaneously: that of selective pressures directly affecting the lean body mass, and that of stress factors operating against a full genotype penetrance in the determining of body size.

A number of studies have, however, shown that tooth size and stature are very poorly correlated (Perzigian, 1981; Wolpoff, 1985). This objection was dismissed on the grounds that '. . . these variables are no longer subject in the same way to the same forces as in the past' (Macchiarelli & Bondioli, 1986, p. 411).

It is the very independence of dental and osseous elements that forms the nexus of another model for negative odontometric trends in the dentition of man. Gloria Y'Edynak analysed jaw and tooth reduction in a series of Mesolithic skeletons from Vlasac in Yugoslavia (Y'Edynak, 1978). Although these people had practised semisedentary fishing and collecting with no evidence of dietary change, there was a marked trend towards jaw size reduction. This was associated with changes in attritional patterns suggestive of a gradual decrease in the biomechanical stresses of mastication. Excavations at the site suggested that these changes were concomitant with alterations in tool materials and also in food preparation and storage techniques. Furthermore, individuals excavated from more recent horizons showed demonstrably higher levels of malocclusion than earlier individuals. The author argued persuasively that changes in food preparation techniques had resulted in an initial phase of jaw size reduction, which in turn resulted in malocclusion, particularly in the anterior teeth. Noting that the greatest dimensional reduction had been found in the MD dimensions of anterior teeth of more recent individuals, she further suggested that the maladjustment between osseous and dental elements had created selection pressures for the reduction in tooth size.

## 5.2.2    *Reduced selection*

In 1967 Brace approached the question of structural reduction of teeth during evolution from a purely theoretical point of view. Arguing that because observed morphology is the end-product of a process of development and growth which is under the direct influence of numerous enzyme chains, Brace postulated that under conditions of relaxed selective pressure, mutations would tend to accumulate, resulting in enzymic inefficiency and hence phenotypic reduction. Brace termed this process the probable mutation effect (PME) and saw in it a possible explanation for dentofacial reduction in man. In a subsequent paper (Brace & Montagu, 1965) a theory was advanced that the anterior teeth in humans had remained large until manufactured foods replaced many of the functions previously accomplished by the incisors. Their line of reasoning was as follows: tool use relaxed the selection pressure for large incisors, rendering these teeth selectively neutral, which in turn resulted in an accumulation of random mutations, and hence (via the PME) culminated in reduction in size of the incisors.

Despite its intuitive attractiveness, Brace's model failed to gain general acceptance. Holloway (1966) noted that the PME required the presence of only 'good' mutations – those that were either nonlethal or did not interfere with the ontogenetic process. Furthermore, Holloway felt that some phenotypic reductions might be explained on the basis of positive selection rather than of neutralism. More critical was the view expressed by Bailit & Friedlaender (1966) that, far from being selectively neutral, the incisors might retain their selective significance as members of the orofacial complex which, in itself, is subject to the process of natural selection. Leaving his critics unanswered, Brace (1967) followed with the suggestion that by the mid-Pleistocene, human incisors had actually increased in size, with a sharp drop in molar size. Brace postulated that the decrease in molar size was due to 'reduced compulsory mastication', while the increase in incisor size would have been related to the increased use of these teeth as all-purpose tools for cutting, tearing and holding. Brace underlined these statements by further positing that he found it less than surprising that the technologically most primitive representatives of modern man – the Australian Aborigines – have the largest teeth in the world. That the hunting and gathering San of southern Africa have among the smallest incisors in man (van Reenen, 1961, 1964) was, however, not mentioned.

If Brace failed to answer his critics, Wolpoff did so with alacrity. The notion that the accumulation of random mutations following reduced selection led to simplification and reduction of a structure, appealed to Wolpoff (1969, 1971). While accepting that structural reduction could not be seen as an inevitable result of the accumulation of mutations, Wolpoff held that its occurrence was a *possible* result. Holloway's objection that the PME was tautological was dismissed with a prediction that, whenever structural reduction was observed, it would be accompanied by increased variability during the period of reduction. In his review of Wolpoff's book (1971), Leutenegger (1972) attacked the foregoing model on two grounds.

First, he felt that the relevance of tooth size as one of the major adaptive mechanisms by means of which individuals meet their environment was overemphasised. The reduction of chewing area throughout the Pleistocene, it was argued, was as much dependent on a reduction in cusps as it was on a decrease in tooth size. For example, a dimensional reduction might conceivably be compensated by the increase of chewing area resulting from an increase in cusps or cuspules. Secondly, Leutenegger (1972) objected to Wolpoff's unfounded inversion of the statement made by Simpson (1953), *inter alia*, that reduction due to direct selection will not be accompanied by an increase in variability. Leutenegger's call for a statistical evaluation of differences in the coefficients of variation that are supposed to result from a relaxation of selection has sadly not been answered to the present day.

Evidence contrary to the application of the PME to human tooth size reduction came from Le Blanc & Black (1974) who investigated dental dimensions in a series of Greek and Turkish populations ranging in time from 7000 BC to AD 1500. Their study showed significant differences in the degree of dental reduction experienced by maxillary and by mandibular teeth, raising the possibility that different mechanisms governed tooth size determination in the different jaws.

Comparisons of the dental dimensions of different age groups within a given population (e.g. juveniles and adults) have been usefully employed in the demonstration of directional selection on tooth size (van Valen & Weiss, 1966). Stimulated by similar work on European cave bears (*Ursus spelaeus*) (Kurtén, 1957) and Middle Miocene horses (*Merychippus primus*) (van Valen, 1963), Perzigian (1975) sought to evaluate selective pressures on human teeth in a protohistoric Arikara Indian population from South Dakota. The tooth sizes of two age groups (juveniles: 6–15 years, adults: 16–20 years) were compared and selection intensity determined using van Valen's graphs. His results clearly demonstrated that younger individuals had smaller and more variable teeth (mean SD juveniles = 0.58; mean SD adults = 0.46). To Perzigian (1975), it appeared that his results indicated directional selection for larger teeth in the presence of increased dental attrition. What he did not realise, however, is that his results may have provided empirical support for Wolpoff's (1969) prediction that structural reduction would be accompanied by increased variability.

In 1976 Brace investigated tooth size reduction in the Orient and concluded that it was a change in cooking techniques rather than a change in diet that had reduced the selective significance of the dentition. By his own admission, however, Brace had failed to provide statistically significant evidence, which tempted him to conclude that '. . . if the reader is not satisfied that this represents a further example of change that is the result of mutations accumulating under relaxed selection, then I offer the challenge to devise a better explanation' (p. 216). Two years later, Williams (1978) published a paper heavily critical of the PME. Citing the work of Kimura & Ohta (1973, 1974) Williams argued that neutral alleles could not have participated in the reduction of hominid phenotypic traits during and after the Pleistocene. The reasons, according to Williams, were that neutral alleles have low mutation rates, experience slow evolution by random genetic drift and have a benign effect on the structure and function of protein.

Undeterred by such criticism, Brace employed the PME to account for tooth size changes in Oceania and Australia (Brace, 1980; Brace & Hinton, 1981). According to his model, two separate waves of genetically different people were responsible for the peopling of the Pacific: an initial Late Pleistocene movement of large-toothed individuals, followed by a second (post-Pleistocene) wave of small-toothed Austronesian-speakers (Figure 5.1). Criticism of this model centred around Brace's use of pooled male and female data (Pietrusewsky, 1980; Brown, 1981; Xiang-Qing, 1981), and around his inference that teeth are little affected by the environment (Brown & Townsend, 1980; Koritzer, 1980; Preston, 1980; Mizoguchi, 1981a). Although Brace has as yet not answered the critics explicitly, he did make use of separate data for the sexes in his most recent analysis of tooth size changes in Japanese (Brace & Nagai, 1982). Brace's earlier studies had focused on human dental reduction since the Pleistocene. From a subsequent analysis of a series of prehistoric and modern groups, he concluded that the PME could be used to explain human tooth size dimorphism. Brace & Ryan (1980) argued that sex differences in tooth size were residues of the sexual dimorphism that was maintained for adaptive reasons during the Middle Pleistocene. Sophistication of food processing techniques during the Late Pleistocene led to the reduction of male and female dental dimensions. However, dental sexual dimorphism was maintained until the hunting of large animals by crude techniques was replaced by small game hunting and an increased utilisation of plant foods. Brace felt that the subsequent reduction in dimorphism represented the action of the PME under conditions of relaxed selection.

Support for the theoretical background of the PME has most recently come from McKee (1984), who used computer generated simulations of the accumulations of mutations in a human gene pool. These simulated genetic changes through time support a polygenic model of the PME as a possible explanation of dental reduction, but much of the work on PME is bedevilled by the absence of any test for evolutionary neutrality. Lack of statistically significant evidence of selectivity is not a demonstration of neutrality.

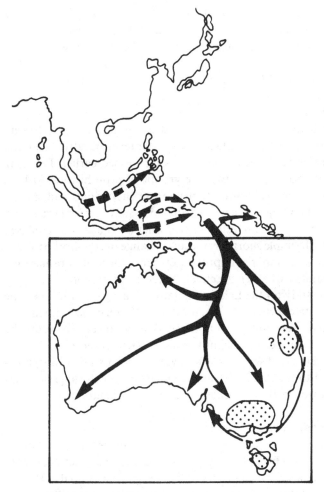

Figure 5.1. Hypothetical routes for the initial Late Pleistocene (interrupted arrows) and the second, post-Peistocene wave (solid arrows), of gene flow into Oceania and Australia (enlarged). Screened patches indicate areas of relatively low degrees of dental reduction. (After Brace, 1980 and Brace & Hinton, 1981.)

Brace, Rosenberg & Hunt (1987) recently examined the changes in human tooth size that have occurred since the Late Pleistocene. Their results showed that, between the onset of the last glaciation (approximately 100 000 years ago) and the end of the Late Pleistocene (approximately 10 000 years ago), human tooth size had reduced at a rate of 1 % for every 2000 years. This rate, however, doubled to 1 % per thousand years from the post-Pleistocene to the present. Brace *et al.* (1987), argued that this doubling in reduction was not the result of dietary change, but rather of

58      Odontometric trends

developments in the realm of food processing. The development and widespread utilisation of pottery after the end of the Pleistocene, together with the use of grinding and milling tools, contributed to a relaxation of Pleistocene levels of selection. Less stringent selection, it was suggested, would operate via the PME to result in a marked decrease in tooth size.

### 5.2.3    Negative secularity
Although the present section has dealt almost exclusively with evolutionary dental reduction, instances of shorter term, negative secularity in human tooth size have been reported. As mentioned earlier, Perzigian (1975) contrasted tooth sizes in two age groups of protohistoric Arikara Indians. His results clearly showed that younger individuals had smaller teeth than adults, a finding which was attributed to selection for larger teeth in the older age group. The question arises whether the author confused selective effects with simple short-term secular change over a short period of time. Older–younger comparisons provide evidence for secular rather than evolutionary changes, whether selectively determined or not.

A more recent study on the Lengua Indians of Paraguay confirmed the existence of a negative secular trend in tooth size. When dental dimensions of older individuals were compared to those of the younger age group, the latter were found to be significantly smaller (Kieser et al. 1985a,b). Canonical and stepwise discriminant analyses isolated statistical significance of this trend to the maxillary postcanine teeth, which suggested that the upper and lower teeth have different dimensional stabilities.

### 5.3    Positive odontometric trends
One of the first authors to document a positive secular trend in the human dentition was Hasund (1966) who observed that the mean MD diameter of teeth in a sample of mediaeval Norwegians was smaller than in contemporary inhabitants of the same area. Similarly, Lunt (1969) found the teeth of mediaeval Danes to be smaller than those of modern Swedes, whilst Ebeling and his co-workers (1973) demonstrated a positive secular trend in the tooth size of Swedish men over the last 160 years. A positive secular trend has also recently been documented in South African Negroes (Kieser, Groeneveld & Cameron, 1987).

If dental reduction is seen as a by-product of a reduction in stature, then one would be hard put not to accept the reverse. It comes as no surprise, therefore, that Ingervall, Lewin & Hedegard (1972) actually found a secular increase in stature and in cranial size in the same population investigated by Ebeling et al. (1973). However, two factors militate against accepting tooth size increase solely as a by-product of secular increase in stature. First, in spite of the claimed significant positive trend in Negro teeth, much evidence points to an absent, neutral or negative secular trend in their stature (Tobias, 1975; Tobias & Netcher 1976, 1977). Secondly, as noted earlier, tooth size and body size seem to be poorly correlated.

In 1968 Garn and his co-workers advanced the suggestion that positive secular trends in the dentition might have their origin in improved nutritional status (Garn *et al.*, 1968c). In fact, a great deal of information had previously been published indicating that tooth calcification may be affected by a number of dietary factors such as fluoride (Cooper & Ludwig, 1965), vitamin A deficiency (Paynter & Grainger, 1956) and various alterations in maternal diet (Holloway, Shaw & Sweeney, 1961; Shaw & Griffiths, 1963). More recently, Corruccini (1984) and Corruccini & Lee (1984) reiterated the classic thesis that, with the westernisation of urban diets, the incidence of malocclusion had increased, especially from maxillary overbite. The latter they attributed to mesiodistal enlargement of the posterior teeth and buccolingual enlargement of the anterior teeth. Evidence for this suggestion may be found in the original paper presented by Garn and associates (1968c): whilst 50% of postcanine teeth showed a significant MD increase, only 16% of the anterior teeth significantly increased in their MD diameters. Further evidence has emerged from data on tooth size enlargement in South African Negroes (Kieser *et al.*, 1987). These authors reported that postcanine teeth displayed the least stability in their MD diameters, whilst the anterior teeth emerged as having the least stable BL diameters.

However, do the above studies really provide unequivocal support for Garn's original suggestion? Perhaps not. The view that a positive secular trend in the dentition has its origins in the improvement of nutritional status may be an oversimplification. Our own studies have shown a positive odontometric trend in a rapidly urbanising Negro population (Kieser *et al.*, 1987). Hence one might have expected that the increased contact with 'more sophisticated' western diets and medicine would have resulted in the observed positive secular trend. The fact is, however, that numerous authors have underlined the nutritional inadequacies of the partially westernised diet of urban Negroes in South Africa. Briefly, the transition from rural to urban living has resulted in recognisable deficiencies in riboflavin, niacin, tryptophan, protein, folic acid, dietary fibre and iron (Walker, 1966, 1982). Obviously the great difficulty in evaluating environmental causes of tooth size modification lies in the multiplicity of environmental modifiers actually or potentially present in the diet.

From the preceding discussion, it is clear that there is an immense amount of information available concerning tooth size changes in various human populations. However, when it comes to causation, the information becomes very inadequate and speculative. Odontometric change may result from change in natural selection pressures or in environmental factors such as diet. However, the fact that we have documented a significant trend in tooth size does not guarantee that its cause lies with either natural selection or environmental effects. In fact, a random or chance process could produce the same results. One such process is the Sewall Wright effect or random genetic drift.

### 5.4     Random genetic drift (Sewall Wright effect)

In *The Origin of Species*, Charles Darwin (1872) stated 'I am convinced that natural selection has been the most important, but not exclusive, means of modification' (p. 4). It has been obvious to many, including Darwin, that natural selection does not determine all the characters distinguishing different subpopulations. Moreover, characters determined by natural selection, and hence adapting the organism to its environment, cannot be used by taxonomists for purposes of classification. Realising this, Darwin noted '. . . we can clearly understand why analogical or adaptive characters, although of the utmost importance to the welfare of the being, are almost valueless to the systematist. For animals, belonging to two most distinct lines of descent, may have become adapted to similar conditions, and thus have assumed a close external resemblance; but such resemblances will not reveal – will rather tend to conceal – their blood-relationship' (Darwin, 1872, p. 374). Thus when considering a trait such as tooth size, we must not be unwilling to grant a role to nonadaptive mechanisms in its determination.

It is the nature, rather than the role, of nonadaptive evolution that has troubled evolutionists since Darwin's day. One of the most enduring and compelling explanations for the evolution of nonadaptive (indifferent) characters remains Sewall Wright's theory of random genetic drift (Wright, 1932, 1982). From his own experience in breeding guinea pigs, Wright concluded that if a population were subdivided into small, partially isolated subgroups, there would be drifting of gene frequencies which would result in nonadaptive or accidental differentiation of these isolated groups. What this means is that while in infinitely large populations gene frequencies tend to remain constant (except for the effects of natural selection), finite populations experience a gradual reduction in variability. This reduced variability is due to drifting in gene frequencies and to random fixation of individual alleles (discussed in greater detail by Provine, 1983).

In odontometric terms, it might be profitable to consider the human population structure of an area on two distinct levels: the level of each of the subpopulations and the level of the total population (i.e. the aggregate of the subpopulations). Suppose we consider two such subpopulations which become isolated in an inaccessible area (so that migration between subpopulations does not have to be considered). They will, at the time of isolation, have the same average tooth size and the same frequencies of the genes controlling it as the total population of the area. After a period of generations of random genetic drift in each of the randomly mating populations, one may have lost an allele for small tooth size, the other an allele for large tooth size. The largest percentage of the remaining genes will remain unfixed or segregating. Of course, once an allele is lost it is lost, or once it is fixed, it remains so (except in the unlikely event of mutation or reintroduction by migration). With the passage of time, more and more of the gene frequencies in the subpopulations will diverge from the original and more and more alleles will be lost or fixed. The end result is that our isolated subpopulations will experience a trend in tooth size. Importantly, the change in tooth size had nothing to do with natural selection. It is a chance, nonadaptive change, due to random genetic drift. Finally, the larger the subpopulation, the longer time will be required for the allelic spread.

## 5.5 Other effects

Random forces may operate independently or even contrary to natural selection. In small, finite populations, alleles with no selective advantage may become fixed by random drift, resulting in nonadaptive differentiation. The question now arises whether random drift could have accounted for the odontometric trends that have been documented in modern human populations. The answer is no.

Although the earliest humans may have lived as small semi-isolated bands of hunter-gatherers, modern societies are increasingly characterised by outbreeding and intermarriage of previously separated clans, nations and ethnic groups. This, together with the great size of the present human population, has unleashed a potential for genetic recombination which is of astronomical proportions. Carson (1986) suggested that no other form of life in the history of the earth has ever undergone such a release of genetic variability. It has to be concluded, therefore, that whilst random drift might be championed as a cause of odontometric trends in early isolated human populations, the causes of modern tooth size variation must be sought elsewhere.

Lewontin (1986) listed three alternatives to adaptive selection. First, he noted that nonselective differentiation might be the result of an interaction of random factors with selection. If genes interact epistatically during development, for instance, there may be different stable outcomes to the same selective force. Which of these outcomes will be realised will depend upon chance fluctuations in gene frequencies. Hence, two subpopulations may become phenotypically different even though they are subjected to the same selective forces. Moreover, Simpson (1953) described random or stochastic processes leading to nonadaptive changes which, though individually unpredictable, may collectively produce a measurable average effect. As Wright (1967) noted, these trends may be indeterminate in causation or direction but determinate in variance.

Secondly, allelic linkage may result in nonselective change. If a certain locus or suite of loci is under selection, alleles at other loci linked to the first, but not under direct selection, will also evolve. Hence, linkage effects or genetic hitch-hiking may result in pseudoselective changes in totally unselected genes. Finally, Lewontin (1986) noted that some genes have multiple developmental effects. Hence, selection on one effect may result in unselected or even counter selective changes in others.

In conclusion, to think that changes in tooth size are purely an outcome of natural selection would be as short-sighted as to say that they are solely nonadaptive. Trends in human tooth size may be transient or permanent. Transient changes are affected by environmental factors such as trace elements, chronic illness, mal- or improved nutrition and so on. More permanent, gene-based, diversity in tooth size is the product of natural selection, together with a number of nonadaptive processes.

# 6 *Sexual dimorphism in tooth size*

## 6.1 Introduction

Sexual dimorphism refers to those differences in size, structure and appearance between males and females of a given species or subspecies at an equal age and, where relevant, during the same season (Thompson, 1964). Although males and females of many species differ markedly by the time they reach maturity, they bear a close resemblance to one another at birth. Prenatally, at two months gestational age, the human foetus is morphologically neither male nor female, dimorphism developing under the influence of androgens between months four and six (Dorner & Kawakami, 1978; Hughes & Lambert, 1984).

As they grow and develop, the sexes of a given species begin to differ in size, plumage, colour, behaviour and form. The task of understanding the meaning and origin of differences between male and female organisms of a given species has occupied the most brilliant biological minds for centuries. For ease of reference, sexual dimorphism may be divided into three levels. At the most fundamental level are the *primary sexual characters* that distinguish the sexes: the testes of males and the ovaries of females. Primary sex characters are essential to the process of sexual reproduction, by means of which new individuals are formed that contain genetic material from more than one source (Margulis & Sagan, 1984).

At the second level are those external features that distinguish males from females. These characters function as releasers of social reactions intimately connected to courtship and mating and are referred to as *secondary sexual characters*. Examples of these abound and range from the bizarre structural elaborations found in the tail feathers of peacocks to often mundane contrasts in colouring between males and females. It has been suggested, for instance, that the large canines found in males of so many animal species function as visual sexual signs of dominance and rank (Portmann, 1967; Thenius, 1970; Herring, 1972).

Sexual dimorphism not only is often a matter of differences in colour or structuring, but it may involve *differences in size* between males and females (the third level of dimorphism). There has been considerable speculation among zoologists as to the origin and significance of sexual dimorphism in size. Some (Storer, 1966; Selander, 1967; Amadon, 1975) have suggested

63

Table 6.1. *Summary multivariate statistics for tooth size differences between male and female caucasoids* (*Kieser et al., 1985e*)

| Tooth group | Mahalanobis' $D^2$ | Wilks' Lambda | $Chi^2$ for covariance homogeneity | Percentage correct classification Males | Females | Variables selected by stepwise discriminant analysis |
|---|---|---|---|---|---|---|
| Maxilla |  |  |  |  |  |  |
| P1, P2, M1, M2 | 1.664* | 0.586* | 53.19* | 88.24 | 87.80 | Step 1 M2 (MD) Step 2 P1 (BL) |
| I1, I1, C | 1.378* | 0.676* | 8.26* | 74.07 | 77.27 | Step 1 C (MD) Step 2 I1 (MD) |
| Mandible |  |  |  |  |  |  |
| P1, P2, M1, M2 | 1.604* | 0.610* | 57.73* | 80.65 | 93.33 | Step 1 P1 (BL) Step 2 P2 (BL) |
| I1, I2, C | 1.414* | 0.664* | 16.01* | 70.91 | 84.62 | Step 1 C (MD) Step 2 I1 (MD) |

* $p < 0.05$.

that a difference in size permits males and females to utilise different ranges of sustenance. Others (Reynolds, 1972; Selander, 1972) have argued that sexual dimorphism in size correlates with mating behaviour. Larger individuals, for example, would enjoy an obvious advantage in the aggressive competition for mates.

The present chapter dwells on sex differences in the tooth sizes of modern humans and on their meaning in human biology.

## 6.2 Extent and patterning of human odontometric sexual dimorphism

In considering the levels and patterning of sexual dimorphism in human tooth size, a fruitful starting point is the measurement of dimorphism. As we have seen in Chapter 4, sex differences in tooth size may simply be evaluated by means of univariate tests such as the Student's *t*-test. For instance, Seipel (1946) and Moorrees (1959) reported the dental dimensions of males to be considerably larger than those of females, the largest differences being found in the canines. Garn *et al.* (1964b) found the sex difference in MD diameter of permanent teeth to be 4% of the combined size, again with dimorphism being greatest for canines and least for incisors. Numerous other studies have re-emphasised the general trend outlined above (e.g. Beresford, 1969; Sanin & Savara, 1971; Lavelle, 1972; Arya *et al.*, 1974; Black, 1978; Harris & Nweeia, 1980a, b; Kieser, 1985; Kieser *et al.*, 1985d, e).

To examine further the notion that males have larger teeth than females, Potter (1972) applied multivariate discriminant methods to human odontometric data. Using similar methods, Ditch & Rose (1972) found sex differences in size to be 93% accurate in determining the sex of a subject, the canines being the largest contributors to sexual dimorphism. Recently, Kieser *et al.* (1985e) employed canonical discriminant analysis on the anterior and postcanine tooth groups and showed significant sex differences in all cases (Table 6.1), with Mahalanobis' $D^2$ distances ranging from 1.378 to 1.664. Significant values of Wilks' lambda (Table 6.1) not only confirmed the statistical significance of the distances between canonical centroids, but also indicated high levels of clustering of data cases around their centroids. Stepwise discriminant analyses identified the MD dimension of maxillary canines and second molars and of the mandibular canines, and the BL dimension of first premolars, as the most significant sex discriminators, providing a range of 70.9% to 93.3% correct classification.

At this juncture another conclusion reached by multivariate analyses becomes relevant. From her study of the results of uni- and multivariate analyses for sex differences in tooth size, Potter (1972) noted a number of differences, which she attributed to intercorrelations among tooth size variables, unaccounted for in univariate analyses. Similarly, Kieser *et al.* (1985e) noted a contrast between the results of univariate and of multivariate sex analyses. Of the highly significant sex differences indicated by univariate *t*-tests, two variables (LM1 – MD and UM1 – BL) were not found by stepwise discriminant analyses to contribute significantly to sex differences. Multivariate analyses in turn selected three variables as highly discriminatory (UI1 – MD, LI1 – MD and LP1 – BL), which univariate tests failed to isolate. Variables selected by both methods were the upper and lower canines (MD and BL) and upper M2(MD), M1(BL) and P1(BL), as well as the BL dimension of the LP2. When these authors analysed the Lengua Indian dentition (1985d), they found a similar discrepancy: of the three variables not contributing to univariate sex discrimination, one dimension (UM3 – MD) was selected by stepwise discriminant analysis to contribute significantly to sex differences. With the exception of the canines, a number of variables with highly significant sex differences from *t*-tests in univariate analyses was found not to contribute significantly to sex differences in multivariate analyses: maxillary I1(MD); M1(BL); M2(MD) and mandibular M2 (MD and BL). From these apparent discrepancies between univariate and multivariate analyses for sex dimorphism, which indicate that some dimensions of different teeth are not independent of each other, Potter *et al.*, (1981) and Kieser *et al.* (1985d,e) concluded that multivariate analyses should be routinely performed in odontometric studies, in addition to the more conventional univariate methods.

In the assessment of sexual dimorphism in tooth size, univariate and multivariate methods have been employed in the analysis of size. The patterning of dimorphism, on the other hand, is possibly best investigated by means of the percentage sexual dimorphism suggested by Garn *et al.* (1967b). The usefulness of this method is clearly illustrated in Table 6.2 which lists percentages and rankings of sexual dimorphism for the MD diameters of a number of Amerindian populations. Harris & Nweeia (1980) noted an unusual lack of sexual dimorphism in mean tooth size of Ticuna Indians and suggested that this might in fact be characteristic of indigenous South Americans. Although the data presented in Table 6.2 do not confirm this suggestion, they do indicate a statistically significant concordance for percentages and rankings of sexual dimorphism between the Lengua, Pima and Ticuna Indians. The percentage dimorphism of the total dentition, expressed as the mean of the MD percentages of all the individual teeth (Perzigian, 1976), yielded a value of 4.1 which was considerably higher than that of either the Pima (1.7) or the Ticuna (1.73), but fell well into the range computed for Ohio Valley Indians (2.6–6.7) by Sciulli *et al.* (1979).

Dimorphism research has only recently begun into factors other than simple tooth size comparisons. One such study is that of Harris & Bailit

Table 6.2. *Percentages and rankings of sexual dimorphism in mandibular and maxillary mesiodistal tooth size of Lengua, Pima and Ticuna Amerindians by Kendall's coefficient of concordance* $W = 0.58$, $p < 0.05$. (*Kieser et al., 1985d.*)

| Tooth | Lengua[a] $\%^c$ | Lengua[a] Rank | Pima[b] $\%^c$ | Pima[b] Rank | Ticuna[b] $\%^c$ | Ticuna[b] Rank |
|-------|------|------|------|------|------|------|
| UI1   | 4.4  | 9    | 1.6  | 7    | 0.5  | 13   |
| UI2   | 3.6  | 5    | 0.0  | 14   | 0.7  | 11   |
| UC    | 7.5  | 1    | 5.3  | 2    | 3.3  | 2    |
| UP1   | 3.5  | 10   | 0.6  | 8    | 0.4  | 14   |
| UP2   | 0.7  | 13   | 0.5  | 10   | 1.1  | 8    |
| UM1   | 1.8  | 12   | 1.7  | 6    | 0.0  | 10   |
| UM2   | 5.1  | 6    | 2.0  | 5    | 2.5  | 4    |
| LI1   | 4.6  | 7    | 0.5  | 9    | 2.0  | 6    |
| LI2   | 5.7  | 4    | 0.1  | 13   | 0.6  | 12   |
| LC    | 7.0  | 2    | 6.3  | 1    | 4.9  | 1    |
| LP1   | 6.1  | 3    | 0.4  | 11   | 1.1  | 8    |
| LP2   | 0.1  | 14   | 2.6  | 12   | 2.0  | 6    |
| LM1   | 3.4  | 11   | 2.7  | 4    | 2.1  | 5    |
| LM2   | 4.5  | 8    | 1.7  | 3    | 3.3  | 2    |

[a] Figures after Garn *et al.* (1967c).
[b] Figures after Harris & Nweeia (1980b).

[c] $100 \left[ \dfrac{\bar{x}m}{\bar{x}f} - 1 \right]$ (Garn *et al.*, 1967b).

(1988) which reports on sex-specific multivariate correlation matrices derived from a staggering 2650 individuals from the Solomon Islands. Principal components analysis disclosed significant sexual dimorphism in the strengths of size intercorrelations. In 61% of all comparisons, and 87% of all significant cases, females exhibited the higher correlation. These results clearly indicate that females have better size-integrated dentitions than males.

## 6.3    Canine size dimorphism

Sexual dimorphisms are probably of multifactorial origin and expression and hence may be assumed to have evolved for various reasons, making the search for a single explanation rather futile (Pickford, 1986). At the simplest level, differences in tooth size between male and female humans reside mainly in the canines. To summarise some of the salient theories that surround canine size dimorphism in other primates, the most traditional of them views canines as weapons which enable males to defend other group members from predators. In the words of Cuvier (1834) '. . . their canine teeth, being longer than the rest, supply them with a weapon we do not possess . . .' (p. 46). The possible relation between canine size and predator protection has been reiterated by Bolwig (1959), de Vore & Hall (1965), Saayman (1971a) and Lauer (1975). A similarly traditional theory explains canine size dimorphism in terms of sexual selection: '. . . when the males are provided with weapons which in females are absent, there can hardly be doubt that these serve for fighting off other males' (Darwin, 1889, p. 502). That large canines in primates might provide a reproductive advantage was noted by Struhsaker (1969), Saayman (1971b), Clutton-Brock & Harvey (1976), Leutenegger & Kelly (1977) and by Harvey, Kavanagh & Clutton-Brock (1978).

A third theory has emerged in recent years: sexual dimorphism in canine size may be seen as a by-product of body size dimorphism. Using body weight as a measure of size, Leutenegger & Kelly (1977) were able to demonstrate that among anthropoid primates there existed a positive correlation between body size dimorphism and canine size dimorphism. Similarly, Lucas (1982) showed a close relationship between canine size and body weight in male primates, a correlation which was absent in females. At a different level, Lucas (1981, 1982) showed that, when mandibular length was used as a measure of size, canine size in males was correlated to facial proportions. The close relationship between canine size and mandibular length in male primates has since been reported in the vervet monkey (*Cercopithecus aethiops*) and the chacma baboon (*Papio ursinus*) (Kieser & Groeneveld, 1987d, 1988a).

Clearly any or all of these theories may apply to human canine size dimorphism. It is difficult to decide which is correct since our canines have become much reduced in size. How can one apply these theories to a species in which canines have clearly not been subjected to the same selective pressures that seem to have applied to other, non-human, primates? The reasons for canine reduction in hominids were listed by Jungers (1978) as follows:

(1) Natural selection fostered canine reduction after the defensive function of these teeth was obviated by weapon manufacture.
(2) There was a shift in endocrine function, which resulted in secondary canine size reduction.
(3) Canines decreased in size to eliminate interlocking and hence to enhance efficient rotary mastication.
(4) There were increased selective pressures to expand the area of incisivation and consequently to accentuate the cutting ability of the new canine-incisor complex.

(5) There was functional disuse of the oral tissues, which resulted from the increased sophistication of culinary skills.

(6) The canines underwent a compensatory reduction to avoid malocclusion in an increasingly restricted jaw.

In spite of their marked reduction in size, the canines of humans have remained the most highly dimorphic teeth in the dental arcade (Moorrees, 1959; Stahle, 1959; Goose, 1963; Schranze, 1964; Garn *et al.*, 1964b, 1965c; Gustafson, 1966; Furuhata & Yamamoto, 1967; Bailit, de Witt & Leigh, 1968; Alvesalo, 1971; Richardson & Malhotra, 1975; Sciulli, 1979; Brace & Nagai, 1982; Axelsson & Kirveskari, 1983; Kieser *et al.*, 1985d, e). Moss (1978) characterised this dimorphism as follows: the absolute size differences in MD and BL diameters generally approximate 0.4–0.5 mm. Neither dimension seemingly demonstrates a greater dimorphism than the other. Finally, in percentage difference, the male mean canine dimensions exceed those of the female by from 3 to 9%.

It may be argued that males have larger canines than females because they have larger bodies than females. That is, canine size dimorphism may be the result of body size dimorphism. Wolpoff (1978a) was the first to study the scaling of canine size to body size in hominids; an analysis from which he concluded that canines were negatively allometric, becoming disproportionately smaller in animals of larger size. Corruccini (1978), however, re-examined Wolpoffs data and concluded the opposite. Wood (1979) conducted a similar analysis, from which he concluded that the canines were in fact positively allometric. Wood & Stack (1980) followed with a more comprehensive allometric analysis of the canine crown area in higher primates, and concluded that larger animals tended to have relatively larger canines. From these analyses, one is tempted to suggest that male humans have larger canines than females, because they have larger bodies.

Three investigations relate to the scaling of canine size in modern humans. In a pooled sample of early Kentucky Indians, Perzigian (1981) noted a negative allometry between canine size and femur length, and Wolpoff (1985) noted negative scaling of canines to stature in Australian Aboriginals (see Chapter 10). The third investigation reported diametrically opposite results: from their analysis of maxillofacial allometry in Negroes, Kieser & Groeneveld (1988g) concluded that the canine base area scaled positively to upper and lower jaw length. Hence it seems that allometry offers little more than confusion in the explanation of canine sexual dimorphism.

It appears that the further we delve into the origins of canine dimorphism, the less sure we can be about our theories and consequently the more we are inclined to seek reassurance through articulating the observations that should guide us. Garn's canine field theory is an example. Garn and his associates (1967b) posited that there is a field of sexual dimorphism that includes the canines and those teeth adjacent to them. Confirmatory evidence was found in a population of Ohio caucasoids and in seven out of eight primate species studied. Further evidence from sibling analyses in humans showed that sexual dimorphism of the teeth adjacent to the canines was more closely correlated with canine size dimorphism than was that of the more remote teeth of the same morphological classes. The validity of Garn's hypothesis has been undermined, however, with the publication of results that are contradictory to these expected patterns (Bailit *et al.*, 1968; Perzigian, 1976; Kieser *et al.*, 1985d, e).

The developing canine may provide a valuable model system for studying the interdependence of skeletal and dental dimensions in the establishment of sexual dimorphism. For example, Anderson & Thompson (1973) noted that sex differentiation often involves permanent tooth size, the rate of maturation and the lengths of certain fingers. Mandibular canines consistently showed greater sex differences in their dimensions and ages of eruption than the other teeth did. The greatest sex difference in dental arch width occurred across the canines, and the fourth fingers in each hand consistently showed the highest levels of dimorphism. Hence it was decided to investigate the strength of these relationships between dental and skeletal dimorphisms. A longitudinal study involving 83 males and 88 females drawn from the Burlington Growth Centre was embarked upon. Theoretically, one would have expected a close correlation between canine width, canine eruption and the length of the fourth finger. Instead, Anderson & Thompson (1973) noted a lack of concordance between these variables, suggesting distinct genetic control for each.

Taken together, these studies suggest that the reasons behind canine size dimorphism are as elusive as when Baron Cuvier first addressed the problem in 1834.

## 6.4    Odontogenetic timing

Mineralisation of the teeth is recorded as being a continuous, progressive and cumulative process with various morphological stages common to all teeth (for a review, see Thompson, Anderson & Popovich, 1975). Although the rate of mineralisation is the same in males and females (Nolla, 1960), the ages at which different stages of mineralisation are reached differ between the sexes (Fanning, 1961; Moorrees, Fanning & Hunt, 1963; Burdi, Garn & Miller, 1970). With the foregoing in mind, Thompson and his co-workers (1975) studied:

(1) the age of mineralisation,
(2) the length of the time intervals between mineralisation stages,
(3) the age order of teeth according to the mineralisation stages.

Their longitudinal study of 121 males and 111 females from the Burlington Growth Centre demonstrated a sex-specific pattern of mineralisation in all teeth but the third molars, maturation being delayed in males. The extent of this delay was found to increase towards the canines, where it was found to be similar to the delay in skeletal maturation. These sex differences manifested themselves before adolescence, and the dental mineralisation showed no adolescent growth spurt. Changes in the sequence of mineralisation were found to result from a delay in the formation of specific teeth, rather than from an increase in the velocity of growth.

Our understanding of dental dimorphism is beset by those unresolved problems relating to the relative roles of genetic and environmental factors during odontogenesis (Moss, 1978). The interplay between these two factors is subtle and ill-understood. There is a lack of knowledge concerning the gene loci implicated in the genetic contribution to sexual dimorphism. Whilst some have implicated the X chromosome, others have argued for the Y chromosome, a combination of both X and Y sites, and finally for polygenic autosomal linkage (see Alvesalo, 1971). In addition, a host of environmental effects seem as important, if not as confusing. From a careful review of the literature, Moss (1978) was led to conclude that dental dimorphism may well reflect some sort of interaction between a variety of genetic and environmental factors, the nature of which are not, as yet, understood. Moss (1978) proposed that the observed dimensional dimorphism could be a result of sex differences in the thickness of the enamel layer. Whatever the factors that combine to produce this measurable dimensional difference, Moss argued, they would act upon the functioning ameloblasts. Citing data presented by Mjor & Hougen (1974), Moss (1978) proposed that neither the pulp cavity nor the dentinal thickness could be held responsible for sexual dimorphism of tooth crown size. The morphology of the human dentino-enamel junction merely represents a generalised and simplified outline of the dental crown. In both sexes the further production of coronal detail is under the control of ameloblasts. Hence the enamel layer itself is the site of dimorphic odontogenesis. It was concluded that the duration of amelogenesis was involved in the production of sexually dimorphic human tooth crowns.

The foregoing discussion reveals another perspective on human odontometric dimorphism – the developmental. Utilising a developmental approach permits one to move beyond the examination of such static data as adult tooth size. Instead, one is able to consider those dynamic processes of ontogeny that underlie the observed phenotypic differences between males and females. This becomes clear when one realises that a given adult end-result may be produced by a number of different developmental processes. Shea (1986) cited the example of a species of animal where the female is half the size of the male. This highly dimorphic end-result could have been produced by the male growing twice as fast as the female for the same length of time, or by the male growing at the same rate as the female but for a period twice as long. Of particular importance is the additional question of whether it is the final difference in adult size, or whether it is the sex-differentiated growth pattern, which is selectively advantageous (see Fedigan, 1982; McNamara, 1982). Although Shea (1983a, 1985a, b) has usefully employed the foregoing approach in his analyses of cranial dimorphism in apes, it still awaits exploitation in the evaluation of human odontometric dimorphism.

## 6.5    The effect of sex chromosomes

It was shown earlier that the effects of the sex chromosomes on dental dimorphism remain controversial. As two representative studies, one might cite that of Lewis & Grainger (1967), who believed that X-linked inheritance governed human tooth size dimorphism, and that of Garn *et al.* (1967a), who asserted that it resided in the Y chromosome. Dental size dimorphism may be attributable to relative concentrations of hormones produced after the differentiation of either male or female gonads, rather than to any direct effect of genes on the sex chromosomes themselves. However, a growing body of evidence points to a direct effect of genes on the sex chromosomes on both osseous and dental structures. Townsend & Alvesalo (1985) recently reported that this evidence has come from two independent lines of investigation. First, normal populations have been studied where observed correlations between related individuals have been compared with theoretical values expected with sex linkage. Although some of these studies have indicated the existence of X-linkage for tooth size and skeletal maturation (Garn & Rohman, 1962; Garn *et al.*, 1965c; Lewis & Grainger, 1967), others have not (Townsend & Brown, 1978a; Potter *et al.*, 1983). In addition, Alvesalo (1971) analysed male and female cousin groups, as well as siblings, and found evidence that both X and Y chromosomes carried genes that may affect tooth size.

The second line of investigation involved individuals with sex chromosome anomalies, where there has been a loss or duplication of chromosomal material. For instance, Tanner *et al.* (1959) demonstrated delayed skeletal maturation in children with Klinefelter's (XXY) and Turner's (XO) syndromes. It was concluded that the Y chromosome produced retardation when one or two X chromosomes were present. In the absence of a Y chromosome, neither one nor two X chromosomes had an effect. Hence it appeared that genes on the Y chromosome were the cause of normal sexual dimorphism in the rate of maturation. Following earlier studies on 47 XYY males (Alvesalo, Osborne & Kari, 1975; Alvesalo & Kari, 1977), Townsend & Alvesalo (1985) contrasted the tooth sizes of 47 XYY males with those of normal controls. Their results indicated a significant increase in the dental dimensions of the 47 XYY males. This suggested that the extra Y chromosome had an effect on tooth size. An interesting additional observation was that the increase in tooth size in the XYY males was not uniform throughout the dentition. In accordance with an earlier study by Alvesalo *et al.* (1975), the canines were least affected. If we bear in mind that canines are the most sexually dimorphic in normal populations, it may be reasoned that they are subject to greater genetic and developmental independence than the other teeth. Although these results appear to validate the theory that the Y chromosome has a direct effect on tooth size, it remains unclear as to whether this effect should be attributed to specific genes on the Y chromosome, or to a more nonspecific heterochromatic effect on cellular activity.

In conclusion, we do not have and are unlikely to have a pure and simple theory for the explanation of dental dimorphism in man. In the words of one of Oscar Wilde's characters – the truth is never pure and never simple.

# 7 *Odontometric variability*

## 7.1 Introduction

In seeking to understand the metrical variability of human teeth, two directions of inquiry seem most plausible. One may start with the tooth as an end-product of odontogenesis and look back, beginning with the embryological evidence, morphodifferentiation and biochemical generalisations, in moving toward an understanding of the genetic basis of tooth formation. Alternatively, one may start with the observed variability and look forward and attempt to formulate mathematical relationships between the variabilities of different teeth. In a completely successful formulation, these two approaches should converge.

The present chapter deals with some aspects of the metrical variation of human teeth. It looks back to the various theories of odontogenesis and looks forward to the patterning of variability. Its failure to synthesise these two approaches should be seen as a call for more concentrated research in this field.

## 7.2 Butler's field theory

Huxley & de Beer (1934) have emphasised perceptively the existence of gradation of morphogenetic substances within the developing embryo. For instance, in *Amblystoma*, the forelimb develops from a discoid zone or field on the side of the body. It is important to note that these authors could not identify a definite area within the field from which the limb always generated. Hence they concluded that inherent in all the regions within the field, was the power of limb formation. They were, however, able to show that limb forming potency was highest in a subcentral region within the field, from which the effect tapered off towards the periphery. Their inevitable conclusion was that morphogenetic substances aggregated in definite presumptive areas within the embryo, from which different organ systems were generated. Each of these areas was termed an embryonic field which, it was argued, acted as a self-contained unitary system.

Credit for the introduction of the embryonic field theory to the process of odontogenesis has to go to Percy Butler. From his review of morphological similarities between adjacent teeth in Zalambdodonts, Butler (1937) concluded that teeth were metameric organs. That is, teeth evolve and develop as part of a system and not as individual units. Drawing attention to the detailed similarity that could be demonstrated for adjacent teeth, Butler further argued that such close correspondence could be attributed only to a common morphogenetic cause acting on more than one tooth germ. In a subsequent paper, Butler (1939) suggested that in addition to the close similarity of adjacent teeth, there was an observable mesiodistal gradation of form along the dental arch. To Butler this meant that in the embryonic jaw there were regions governed by the actions of morphogenetic substances, within which regions tooth rudiments would develop according to the surrounding morphogen. Each field, he argued, had a mesiodistal axis within which there was a definite gradation of morphogenetic concentration. Tooth germs forming at different locations were acted upon by three morphogens corresponding to incisor, canine and molar districts. Because the highest concentration of molar morphogen was to be found in the region of the M1, this tooth could be considered to be a polar or most stable tooth within the molar field. A progressive diminution of morphogenetic concentration in either direction, it was suggested, resulted in greater variability of form towards the ends of the molar field. To explain why the posterior molars were dissimilar to the more anterior premolars, Butler (1939) suggested that tooth germs developing at different positions along the mesiodistal axis of the dental arcade responded differently to the same morphogen.

Fully understanding the usefulness of Butler's field theory to explain the patterns of dental morphodifferentiation in mammals, Dahlberg (1945) sought to apply this theory to the human condition. Describing a field as a sphere of influence, Dahlberg identified four such morphogenetic fields in each jaw: incisor, canine, premolar and molar. It was argued that within each field there were stable or 'key' teeth which would retain structures and traits even when the latter were missing from the more peripheral or 'variable' teeth (Dahlberg, 1986). The polar teeth identified by Dahlberg were the central incisor, the canine, the first premolar and the first molar. These teeth, he suggested, were the more conservative in form and size and hence most closely approximated the ancestral form of the tooth class.

Fifty years earlier than Dahlberg's original paper, William Bateson (1894) published an extensive review on the patterns of variation in natural populations. One of Bateson's conclusions, namely that the sites of greatest variability tended to be at the beginning and end of a series of segmentally repeated structures, had a decisive influence on Dahlberg's thinking. In line with Bateson's conclusion, he suggested that the distal or more posterior members of each tooth group in man could be expected to be the most variable.

One argument against Butler's field theory was that it failed to explain the discordant variation of dental characters along the tooth row. In an attempt to explain this phenomenon, van Valen (1970) proposed what he termed a gradient prepattern model. Briefly, van Valen suggested that all the presumptive tooth buds have identical prepatterns, different parts of which are activated to different degrees by gradients of evocators. It was argued further that any tooth primordium could be regarded as competent to produce any tooth (and, for that matter, a variety of tooth forms that do not normally appear). In agreement with Butler, van Valen saw the end-product as determined by its unique position within a set of morphogenetic gradients. Contrary to Butler, he argued that the response generated by the prepattern need not necessarily be proportional to the concentration of morphogenetic substance; whilst morphogenetic gradients may explain the existence and direction of differences among teeth, the magnitude of these differences depended entirely on the prepattern.

## 7.3     Osborn's clonal theory
It is axiomatic to both the field and prepattern models that all the dental primordia should be identical – differences between teeth are provoked by outside influences. One who has been particularly unhappy over this emphasis on the role of field substances is J. W. Osborn. His first important work was on the embryology of tooth replacement in reptiles, which stimulated his interest in the problem of dental morphogenesis and particularly the early development of teeth. As far as Butler and van Valen's reliance on field substances was concerned, Osborn objected to the concepts of primordial equivalence and external odontogenetic control (Osborn, 1970, 1971, 1973, 1975, 1978; Osborn & Ten Cate, 1983). Was it reasonable he asked, to consider the primordium of a human deciduous incisor (which develops at 8 weeks *in utero*), to be identical to the primordium of a third molar tooth which develops at 8 years of age? In direct opposition to the field model, Osborn proposed a clonal model which relied on the self-determination of primordia.

According to Osborn, human teeth are derived from three clones of identical ectomesenchymal cells: the incisor, canine and molar clones. Consider, for example, the molar clone situated in the region of the first deciduous molar. Together with the overlying dental lamina, which it has induced into odontogenetic cooperation, the clone expands posteriorly. As soon as the growing clone has reached a critical size, a tooth bud is initiated at its centre. Further budding is prevented by a zone of inhibition that surrounds the tooth bud. Continued growth of the clonal progress zone beyond the zone of inhibition will result in the initiation of a further tooth bud, surrounded by its own zone of inhibition. As and when space becomes available within the posteriorly expanding clone, new primordia will be initiated. Because the cells giving rise to later developing primordia will have divided more than those of earlier primordia, there is a gradient of cell ancestry of successive tooth buds. Osborn (1978) ascribed the greater variability of later forming teeth to this gradient, rather than to the gradient in field substance favoured by Butler.

## 7.4   Epigenetic canalisation

Morphogenesis concerns the set of developmental sequences by means of which different tissues, organs and ultimately the organism as a whole are generated. It tells the story of how information encoded in the DNA of a single-celled zygote is elaborated into a highly complex multicellular adult. The viewpoint that the size of a tooth is specified by its genes provides too narrow a context for the understanding of those ontogenetic and phylogenetic processes involved in the formation of teeth. As we have seen, ectomesenchymal cells have a certain competence which, together with some initial positional identity, allows them to commit themselves to partial differentiation. Thereafter, the tooth buds undergo progressive narrowing of competence or developmental patterning until the adult crown size and shape is reached.

This visualisation of the odontogenetic process had its beginnings in the works of Waddington (1940, 1957) and Schmalhausen (1949) who developed the idea of canalisation or autoregulation of development. These concepts may best be visualised by representing the developing cell as a train standing on the single track which gives it access to a marshalling yard. Soon after entering the yard, the track diverges at a number of switches to reach a given number of unique end-points. At the entrance to the marshalling yard, the train can theoretically reach any of these end-points. However, once an initial switch has been made, only a given subset of railway endings remains possible. Thus the train starts in a totipotential state, after which it becomes increasingly restricted by determinative events at each switch. This process may be referred to as canalisation or autoregulation: once the train is on the correct track, canalisation will keep it there, even in the face of environmental disturbances.

When these general properties are applied to the process of tooth formation, the omnipotent ectomesenchymal cells can be seen to become more and more restricted by determinative events during odontogenesis. The canalisation of tooth formation involves a 'black box' of interactions that occupy the middle ground between genetic input and phenotypic output; between DNA and tooth. These mechanisms are highly integrated developmental processes that are distinct from the genome and that produce detailed phenotypes which are neither rigidly nor explicitly

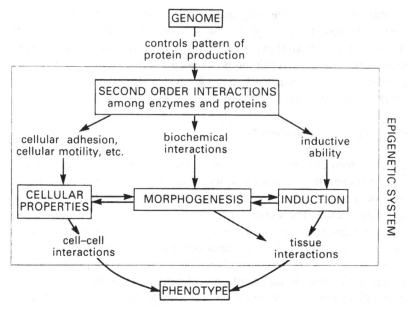

Figure 7.1. Alberch's developmental flow-chart (Alberch, 1982).

encoded in the DNA (Katz, 1982). The 'black box' responsible for the depth of canalisation may best be characterised by the hierarchical organisation of development presented by Alberch. Beginning in 1980, Alberch published a series of incisive papers on developmental biology and evolutionary genetics. These researches dealt with the effects of ontogenesis on morphological diversification (Alberch, 1980; Alberch & Alberch, 1982) and with epigenetic constraints during development (Alberch, 1982). One of his most important suggestions was that developmental processes could be characterised by a hierarchy of three interacting levels of organisation (Figure 7.1). The first level consists of the highly integrated genome which controls the patterns of protein production. Level two involves second order interactions among enzymes and proteins which govern cellular properties, morphogenesis and inductive processes. Finally the third level consists of the establishment of the final phenotype. Variability of the end-product may have its origin in the genome or in environmental effects during the process of odontogenesis. Both of these sources of variation are resisted by canalisation which is an epigenetic (level two) system. Hence phenotypic variability may be seen as a measure of the canalising ability of the epigenetic system, or as a measure of the genetic or environmental disturbances that may have occurred during tooth formation.

In conclusion, it transpires that there are three major hypotheses that

offer explanations for odontometric variability. Variation in dental dimensions may be related to a diminution of field substance (*vide* Butler) or to the
relative age of the ectomesenchyme (Osborn) or to the depth of epigenetic
canalisation (Waddington). The latter half of this chapter will examine each
of these theories in the light of the observed patterns of metrical variability
in the human dentition. Additionally it will probe the statistical associations between variability patterns, in the light of some other hypotheses
that have been advanced.

## 7.5    Patterns of variability

As noted earlier, Dahlberg (1945) applied the field concept to the
process of odontogenesis in humans, identifying four morphogenetic fields
in each half-jaw: incisor, canine, premolar and molar. Describing an
odontogenetic field as a sphere of influence, he suggested further that its
effect would diminish with increased distance from the most stable or polar
tooth within each field. It was argued that in each field teeth with the lowest
variability were at or nearest the centre of morphogenetic influence,
whereas more variable teeth were thought to reside at or nearer the
extremities of each field. Dahlberg (1945) identified the most stable or polar
teeth in humans as the maxillary I1, C, P1 and M1, and the mandibular 12,
C, P1 and M1. In contrast, Osborn (1973, 1978) proposed an alternative
odontogenetic mechanism which does not require the assumption of
controlling regions under the influence of morphogens. Essentially, Osborn
believed that differences in teeth within the arcade reflected differences in
the mesenchyme. From three types of mesenchyme there differentiated
three tooth primordia, each of which being the progenitor from which all
elements of that tooth class subsequently develop.

Many investigators have suggested results generally supportive of
Butler's field theory (van Valen, 1961; Garn *et al.*, 1967a, b; Potter *et al.*,
1968; Suarez & Bernor, 1972; Suarez & Williams, 1973; Alvesalo &
Tigerstedt, 1974; Henderson & Greene, 1975; Lombardi, 1975; Harris &
Nweeia, 1980b; Kieser 1985), but there are those who have questioned its
validity (Biggerstaff, 1970; Kieser, 1986; Kieser *et al.*, 1985d, e). One way of
evaluating the patterns of variability of teeth is by looking at variability
profiles. These consist of arrays of values of coefficients of variation
($CV = 100SD/\bar{x}$) calculated for each suite of measurements. Figures 7.2 and
7.3 contrast the variability profiles for MD dimensions in male and female
Negroes, caucasoids and Lengua Indians. A concordant pattern is seen to
emerge, with first and second molars, canines and first incisors tending to be
the least variable teeth.

In the mandible, however, there is a marked tendency for second incisors
to be less variable than first incisors (Figure 7.3). Noting a reversal of

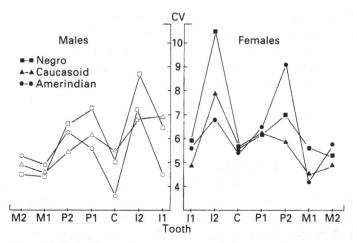

Figure 7.2. Variability profiles for maxillary MD dimensions in male and female Negroes, caucasoids and Lengua Indians. (Kieser & Groeneveld, 1988b.)

expected variability patterns in the second molars of the prosimian primates, the *Indriidae*, Gingerich & Schoeninger (1979) suggested that metric variability might be inversely proportional to occlusal complexity. Their hypothesis rested heavily on the assumption that the maintenance of an accurate occlusal relation would have a higher selective value for cheek teeth than for canines: 'Teeth with simple occlusion (canines) can be relatively more variable in size and form than teeth with complex occlusion (cheek teeth) and still occlude at the same level' (p. 462). However, Figures 7.2 and 7.3 show that MD dimensions of incisors were the most variable, and those of the molar groups the most stable. Although these findings appear to support the occlusal complexity theory, it was found also that canines, which have a simple occlusion, tended to be less variable than the more complex premolars (the sole exception being mandibular premolars in males). A comparison of BL variability similarly revealed little support for Gingerich & Schoeninger's hypothesis: lower molars in females were more variable than canines, upper premolars in males were more variable than canines and, finally, lower premolars in females were less stable than incisors (Kieser & Groeneveld, 1988d). Hence the data presented here are not consistent with the view that teeth with complex occlusions are less variable than those with simple occlusions.

As another interesting result of their investigation, Kieser & Groeneveld (1988b) reported that there was evidence for a negative correlation between tooth size and variability. Pengilly (1984) reported the same result in his analysis of dimensional variability in the arctic fox (*Alopex lagopus*) and explained it on the basis of a morphological rather than a functional

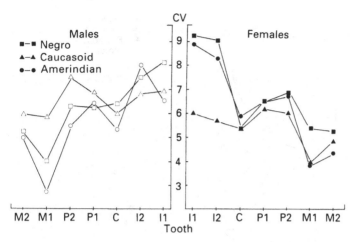

Figure 7.3. Variability profiles for mandibular MD dimensions in male and female Negroes, caucasoids and Lengua Indians. (Kieser and Groeneveld, 1988b.)

hypothesis. Briefly, Pengilly (1984) reasoned that developmental factors might be more important than proximate functional factors in determining the shape of the dental character variability profiles. From his analysis, Pengilly concluded that rather than reflecting strong selection for optimal occlusal precision, the high CV's of large cheek teeth may simply be a consequence of development. Kieser & Groeneveld's (1988b) initial approach to this hypothesis was to regress variability on tooth size. This presented them with the expected negative correlation mentioned earlier. However, the clustering pattern within the regression showed that morphogenetic tooth classes generally formed distinct clusters (Figures 7.4 and 7.5). Comparison with the results of a previous investigation of variability patterns in *Cercopithecus aethiops* and *Papio ursinus* (Kieser & Groeneveld, 1987a,b), where similar variability cluster was reported, showed that morphogenetic clusters had different patterns in the three species, prompting the suggestion that such patterns might be useful in primate taxonomy.

## 7.6    Mizoguchi's group variation theory

The assumption that, in addition to acting on single teeth, selection pressures may act also on groups of teeth, underlies much dentomorphological and human evolutionary thinking. Such theories are implicitly based on the assumption that there are certain groups of teeth that exist as units within the dental arch. It is this argument that has been used for concepts such as the compensatory interactions between tooth classes, differential reduction of tooth size during hominid evolution and the differential

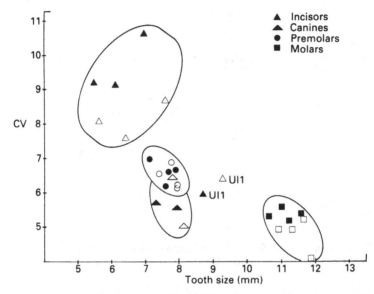

Figure 7.4. Bivariate regression of CV against mean MD tooth size in Negroes (solid symbols = females). (Kieser & Groeneveld, 1988b.)

variability of the four morphological tooth classes. Despite an abundance of theoretical considerations, there is limited experimental evidence for intragroup covariance in human teeth.

In one of the earliest studies of group variation in human dentitions, Lombardi (1975) inferred that four joint factors accounted for 69% of the total phenotypic variance of the MD diameters. Lombardi's inference was subsequently accepted by some as further evidence for observed independent influences acting on the four morphological tooth classes (Hanihara, 1976, 1977). Much of our understanding of group variation of human teeth, however, hinges on the work of Mizoguchi (1977, 1980, 1981b, 1983b). Mizoguchi's treatment of dental variation, first from a genetic point of view and then, considerably expanded, from an intragroup covariance viewpoint, constituted a milestone in modern odontometric research.

Mizoguchi's original study (1977) was more a methodological investigation into the relative importance of genetic and environmental effects on tooth size, than a substantive advocacy of group variability. Heritability estimates of MD dimensions in 191 monozygotic and 75 dizygotic twin pairs were lowest in UC, UP1, LI2, LP1 and LP2. With the exception of the LP2, it was reasoned, each of these teeth could be accepted as the most stable tooth within its respective tooth class. In his attempt to explain these low heritabilities, Mizoguchi thought it appropriate to invoke Falconer's (1967) suggestion that characters with low heritabilities tended to be closely

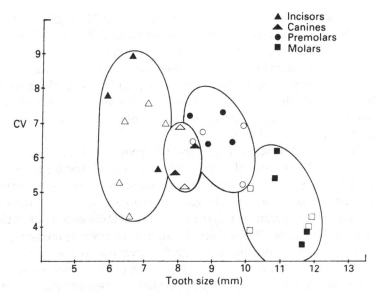

Figure 7.5. Bivariate regression of CV against mean BL tooth size in Negroes (solid symbols = females). (Kieser & Groeneveld, 1988b).

associated with biological fitness. He concluded that the teeth mentioned above, which had relatively low heritabilities, must also have been stable during human evolution.

In his second paper, Mizoguchi (1980) wrote cautiously about the need for a separation of genetic and environmental effects on the dental system. His use of exploratory maximum likelihood factor analysis on MD data from 272 individuals indicated that there were four major genetic factors affecting tooth size. However, his results indicated also that there were two major environmental factors at work, one before and the other during the period of calcification.

Mizoguchi's third paper (1981b) specifically explored the likelihood that there might be tooth groups within the dental arcades that function as units, and which collectively have greater fitness significance than individual teeth. Canonical analysis of MD measurements of 51 males and 52 females revealed that the anterior dentition dimensions (I1, I2, C) were inversely proportional to those of the postcanine dentition (P1, P2, M1 and M2). To explain the independence of anterior and postcanine teeth, he relied in part upon the evidence that, from the Neanderthal stage to the present, the amount of human dental reduction has been greater in the anterior teeth (Brace, 1967, 1979; Sakura, 1970). Mizoguchi stressed that this did not necessarily imply that the anterior teeth were of lesser adaptive significance than the posterior teeth. In fact, he noted that attrition during the Late

Pleistocene was greater in the anterior teeth (Smith, 1977), suggesting that these teeth had a large degree of functional significance. Rather, differential reduction could be attributed to the somatic budget effect. Mizoguchi felt that the inverse relation between these two tooth groups was unlikely to be a reflection of compensatory interaction, more plausibly it could simply be ascribed to independent variability of the units. It was concluded that two primary functional units rather than four morphological classes characterised the human dental arch.

Fascinated by Mizoguchi's concept of group variation, Kieser & Groeneveld (1987c) examined intra-arcade MD and BL tooth size correlations in a sample of 125 children with ideal occlusions. To compensate for arcade length variation, a series of regression analyses was conducted, each MD dimension being regressed on the sum of the MD dimensions of all the teeth within the arcade. This enabled the authors to evaluate intra-arcade tooth size correlates by means of partial Pearsonian coefficients of correlation. Their study demonstrated that when tooth size was corrected for arcade length, the dentition may be divided into two sections that are negatively correlated: an anterior segment consisting of the incisors and canines, and a posterior segment consisting of premolars and molars. Accordingly, when teeth within the posterior segment are larger than expected (for that particular arcade length), then teeth of the anterior segment will be smaller than expected, and *vice versa*.

Early attempts at relating hominid jaw size reduction to a decrease in tooth size recognised only reduction of the posterior teeth (Robinson, 1954), or the most posterior members of each tooth class (Brace, 1967; Sofaer, Bailit & MacLean, 1971; Sofaer, 1973). It was later recognised that anterior and posterior teeth may in fact respond to different selective pressures (Wallace, 1978; Wolpoff; 1978b; Trinkaus, 1983). Moreover, insectivorous primates have larger postcanine teeth than frugivorous or folivorous primates (Hylander, 1975; Kay, 1975a,b; Goldstein, Post & Melnick, 1978; Pirie, 1978; Gingerich & Smith, 1985). Clearly, the recognition of arcade size as having a major effect on or being affected by tooth size requires some reassessment of the theories for hominid dental reduction. For instance, canine and cheek teeth may not have evolved *pari passu* (Wallace, 1978). Nor should canine reduction in robust australopithecines (*Australopithecus robustus*) fossils be seen as necessarily or exclusively the result of selection against an allometric trend – see Chapter 11. It may even be tentatively suggested that the anterior tooth size reduction in *A. robustus* and *A. boisei* might instead be viewed as an ineluctable compensatory reaction to postcanine molarisation or enlargement.

## 7.7 Variability and development

Indicative of the rigour of Mizoguchi's research is the fact that he uncovered a possible developmental cause for metrical variation, whilst investigating the genetics of tooth size determination. Factor analysis of the MD dimensions of 272 monozygous twin pairs pointed to both genetic and environmental causes (section 7.6). Mizoguchi (1980) tentatively postulated that a crucial role might be ascribed to the developmental period prior to calcification. In a later paper, Mizoguchi (1983b) was able to show that the three least variable maxillary teeth (M1, C and I1) also had the shortest precalcific periods. Conversely, the more variable LI1 has a shorter soft tissue stage of development than the more stable LI2. Further support for Mizoguchi's postulate has since come from the reported levels of concordance between the durations of soft tissue stages and the corresponding CV's in negroid and caucasoid teeth (Kieser, 1986; & Groeneveld, 1988b).

The importance of developmental factors to metrical variability has recently been confirmed in three different investigations. Harzer (1987) studied 54 twin pairs and their families to obtain more information on the inheritance of tooth size. More specifically, he was interested in the contributions of genetic differences to the variability of the anterior teeth. The strategy he used in constructing his hypothetical model for the control of anterior tooth size had the following elements: first, heritability estimates were computed for both MD and BL diameters of incisors, canines and premolars. These were found to be highest in the incisor MD diameter, and tapered off towards the premolars. Secondly, heritability was estimated for the summed MD diameters of each tooth group. This showed greater heritability values for each group than for individual teeth. Moreover, the genetic contribution to size variation was found to be considerably higher in the incisor as compared to the premolar region. The final step was to compare these estimates with those calculated for the alveolar bone within the same region.

Harzer (1987) concluded that genetic control over the size of tooth groups was greater than over individual teeth. In contrast to the mesiodistal decline of genetic influence on tooth size, the genetic contribution to the size of the surrounding bone appeared to be higher in the canine region and lower in the premaxillary region. However, Harzer posited that the size of the incisors was determined during a phase of intense bony growth, whereas canines and premolars were formed during periods of decelerating growth. Moreover, Biggerstaff (1979) had suggested that the control mechanisms of tooth formation would be influenced equally by time, environmental feedback and tissue interactions. Accordingly, Harzer (1987) proposed that tooth size variability was affected by a combination of the genetic control of tooth and surrounding bone size and by the genetic control of jaw growth. Further, expansionary potential for the developing tooth germ seemed higher in a BL than the MD direction, from which Harzer concluded that the MD heritability would be reduced.

There are several grounds for criticising the above outline. Perhaps the most significant is that Harzer (1987) failed to reveal the heritability values for alveolar bone or, for that matter, how these were determined. The important point that tooth germs and surrounding bone are genetically interdependent is contradicted by evidence presented on the allometric relations between tooth size and jaw size (see Chapter 10).

Independent factors operating on the MD and BL dimensions of anterior teeth have been the focus of a recent provocative paper by Townsend (1987); provocative, because the author examined crown size variation from a novel point of view. Townsend analysed tooth size correlations in 272 individuals with Down's syndrome, using 321 normal subjects as controls. The reason why Townsend (1987) used individuals with Down's syndrome was that they are more variable in many phenotypic features, which was thought to be the result of a decrease in developmental homeostasis. It was suspected that interdental correlations might be consistently lower in Down's subjects, which would point to a reduction in odontogenetic coordination. In addition, he wanted to find out whether there were changes in correlation patterns that might suggest modifications of regional controls within the dentition. More specifically Townsend's investigation was aimed at the apparent independence of the MD and BL diameters in anterior teeth (Potter *et al.*, 1968; Lombardi, 1975; Townsend & Brown, 1979b; Kolakowski & Bailit, 1981a).

The final size and shape of a tooth is the product of the finely coordinated onset and rate of cellular proliferation within the tooth germ, and the onset, rate and duration of mineralisation during dentinogenesis and amelogenesis. The interplay among these factors is subtle. On the one hand, dentinogenesis in the anterior teeth commences on the incisal edge region and progresses cervically. Hence the basic maximum MD diameter is determined relatively early during odontogenesis. However, the maximum BL diameter, located near the cervical region, is one of the last parts of the crown to be determined. On the other hand, in the postcanine teeth, the MD and BL diameters are situated at about the same occlusocervical levels, hence it may be assumed that these are determined at similar times during crown development. It became apparent to Townsend (1987) that the relative independence of the two diameters in anterior teeth might simply reflect the different times at which they are determined. From his analysis of Down's subjects, however, it was found that the correlations between the diameters had actually increased. A previous study had led Townsend (1983a) to conclude that there was a tendency towards reduction in mitotic activity of developing tooth germs in Down's syndrome. Accordingly, it might not be unreasonable to expect that the later forming parts would be more likely to be distorted than the earlier forming regions. Hence the progression of calcification from the incisal edge towards the cingulum in the anterior teeth of Down's subjects would tend to determine the maximum BL dimension before it had attained its fullest expression.

The foregoing makes it apparent how easy it is to fall into conceptual confusion and also that such confusion is avoidable. The apparent independence of MD and BL dimensional variation of anterior teeth, as opposed to that of posterior teeth, now emerges as a reflection of the differences in the time at which these dimensions are determined during odontogenesis.

This brings us to the third publication to be discussed here. Kieser & Groeneveld (1987c) investigated the association between arcade length and the dimensional stability of teeth. Dimensional sensitivity of individual teeth to arcade length was calculated as follows:

$$\beta = \frac{\sum_{i=1}^{n} (A_i - \bar{A})(M_i - \bar{M})}{\sum_{i=1}^{n} (A_i - \bar{A})^2}$$

where $n$ = sample size, $i$ = $i^{th}$ tooth, $A$ = arcade length ($\Sigma$MD), $M$ = the dimension under consideration, and $\beta$ = the regression coefficient of the dimension on arcade length.

In addition to $\beta$, the proportion of variation in tooth size that may be attributed to variation in arcade length was calculated as $R^2$ – the coefficient of determination. When these values were plotted together with the CVs of MD dimensions, a number of interesting patterns emerged (Figure 7.6). This graph clearly demonstrates the close pattern of concordance between $R^2$ and $\beta$. More intriguing perhaps is the relationship between $\beta$ and the CVs: teeth with low values of $\beta$ clearly tend to have high CVs. High values of $\beta$, on the other hand, are associated with lower CVs, the only exception being LI1(MD). Hence it appears that a high degree of arcade size dependence would confer a high degree of dimensional stability to a tooth. The converse, of course, also holds: low arcade size dependence tends to be associated with high dimensional variability.

## 7.8    Conclusion

The reason why the Butler–Osborn–Waddington debate is so difficult to settle should now be clear; the theories of the last two scholars are not falsifiable using variability data. Whilst a reversal of expected

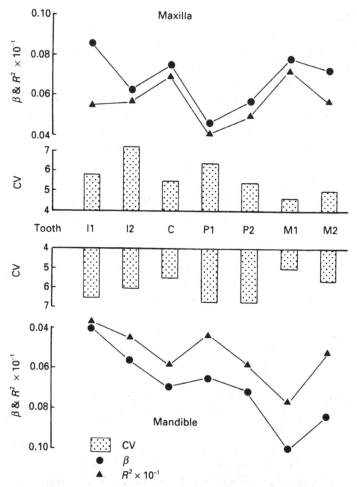

Figure 7.6. The relationship between arcade size dependence ($\beta$), the coefficient of determination ($R^2$) and the dimensional coefficients of variation (CV) for MD dimensions of maxillary and mandibular teeth. (Kieser & Groeneveld, 1987c.)

variabilities might be seen as evidence against Butler's concept, it does not necessarily provide unequivocal support for Osborn or Waddington. Dental dimensional variability emerges as a complex phenomenon and will probably require a complex synthesis of ideas for its explanation. At present it remains difficult to draw a clear picture from so many hypotheses: Waddington's canalisation, Butler's fields, Osborn's clones, Pengilly's functional relations and Gingerich's occlusal complexity. The honest answer at the moment is that we do not know the exact cause for the observed patterning of odontometric variability in man.

# 8  Compensatory tooth size interactions

## 8.1    Introduction

Hominid dentofacial reduction has been a topic of major research during the past hundred years or more. The problems posed by the differential reduction of tooth size and jaw size are complex and raise issues of ontogenetic, phylogenetic and clinical significance. Although it is sometimes asserted that the challenges posed by phylogenetic facial shortening are difficult to relate to empirical orthodontic research, there are in fact a number of points of contact between the two fields of investigation. The object of this chapter is to review one such contact point: that of compensatory interactions between adjacent teeth, in the face of spatial restriction, first by considering the variability of the upper incisors, and then by considering Sofaer's model of compensatory tooth size interaction.

## 8.2    Upper incisor interactions

The theory that human upper lateral incisors, like wisdom teeth, are particularly susceptible to reduction in size, grew out of early attempts to reconcile the progressive shortening of the jaws with a concomitant but less dramatic reduction in tooth size (e.g. Rosenberg, 1895; Wiedersheim, 1895; Rose, 1906; Hrdlička, 1911). This traditional picture was, however, shattered by Ashley-Montagu (1940) who argued that if the upper and lower jaws had undergone an equal reduction there should be an equal reflection of this in the teeth of both jaws. Accordingly, the mandibular incisors would be expected to be affected at least as often as those of the upper jaw.

Ashley-Montagu sought to demonstrate that upper lateral incisor variability in man was related to premaxillary rather than dental reduction. He believed that the relative stability of the upper central incisors and canines as compared with the lateral incisors pointed to a developmental instability of the bony elements within this region. He argued that excessive posterolateral reduction of the premaxilla would result in the constriction of the dental lamina, with a resultant vestigialisation of the lateral incisors. In other words, progressive evolutionary reduction of the upper jaw would have involved primarily the bony diastema posterior to the second incisor. Thus variability of the lateral incisors should not be conceived as an indication of their decreased significance, but rather as a reflection of the diminution of the bony element that surrounds them.

Associations between third molar reduction and other dental variabilities may also provide some information applicable to lateral incisor reduction. Garn and his co-workers (1961, 1963a, b, c, 1964a) studied third molar agenesis and found it to be correlated with delayed initiation of tooth formation in posterior teeth and also with a general reduction in tooth size. Evidence for a similar association between upper lateral incisor reduction and overall MD size reduction of the other teeth also came from Hanihara, Masuda & Tanaka (1965) and Hanihara (1970).

If a given early developing tooth is larger than average, then will those later developing teeth next to it be smaller or absent? This was the question posed by Sofaer and his co-workers (1971a, b) when they examined developmental interactions within the dentition. More specifically, they wanted to know how central incisors would react to missing or peg-shaped lateral incisors. Their investigation showed that peg-shaped laterals were associated with small centrals, whereas missing laterals tended to be accompanied by larger than normal central incisors. In addition, when a lateral incisor was missing on one side, not only were *both* central incisors larger than normal, but the central incisor adjacent to the missing lateral was significantly larger than its antimere. Hence both central upper incisors had reacted to the missing tooth by a compensatory increase in size, the effect being more marked in the ipsilateral central incisor. However, where there was a peg-shaped lateral incisor on one side, the central incisors tended to be smaller than average. Sofaer *et al.* (1971a, b) thus concluded that interactions do occur between adjacent upper incisors, but that compensatory increase in size occurs only if the requirements necessary for growth are not limited. On the other hand, peg-shaped lateral incisors, which were associated with smaller central incisors, were indicative of poor developmental conditions. A similar difference between dentitions with size reduction of the UI2 and those with agenesis of the UI2 was later reported by Le Bot & Salmon (1977) and by Le Bot, Gueguen & Salmon (1980).

## 8.3    Compensatory tooth size interactions

Considering how the hominid face has evolved since the Late Pleistocene, it is obvious that the maxillofacial complex has undergone a dramatic size reduction. Yet the accompanying tooth size reduction has not kept pace with it. This statement neglects the hypothesis that natural selection has operated to maintain a harmonious size relationship of tooth to jaw by tending to eliminate those genotypes that result in teeth too large for the changing skeletal system (Sofaer *et al.*, 1971a, b; Sofaer, 1973). From their review of post-Pleistocene dental reduction in humans, Sofaer and his associates identified two basic trends. First, the cheek teeth have undergone greater reduction than the anterior teeth. Secondly, the more distal members of each morphological class (incisors, premolars and molars) have been reduced more than their mesial neighbours. Sofaer then asked the question: may the differential jaw and tooth size reduction that has taken place during hominid evolution be accounted for in terms of spatial restriction and developmental timing?

He constructed a model which rested on the following assumptions: first, selection was taken to act on the combined phenotype of each morphological tooth group rather than on individual teeth. Secondly, interactions between adjacent teeth would ensure that later forming teeth would compensate for size deviations from the norm in earlier developing neighbours. Thirdly, early developing teeth experience less spatial restriction than later forming teeth. The relative difference between the local levels of environmental restriction experienced by early and late developing teeth would thus allow a more complete realisation of the genetic size potential of the earlier developing tooth. Sofaer then sought to demonstrate a negative correlation between the sizes of adjacent teeth within a given morphological class, which might be indicative of a compensatory tooth size interaction. Finally, the degree of resemblance between relatives was measured by the intraclass correlation for each measurement among sibling and parent–offspring pairs. Comparisons of different teeth with respect to the degree of resemblance between relatives would then provide an estimation of differences in the extent of the genetic control over tooth size between teeth.

From their analysis of 117 parent–offspring pairs and 146 sibling pairs of two Melanesian tribes (Nasioi and Kwaio), Sofaer and his co-workers concluded the following: since skeletal changes have led the way, the teeth of an individual would always tend to be genetically too large for the jaw in which they develop. Environmental deviations, to which the last tooth to develop will be the most susceptible, would tend to be overwhelmingly negative due to the restriction of space. An inability to respond to this local environmental challenge would result in disharmony between tooth and jaw size. Hence selection would operate against those genotypes that failed to produce compensatory interactions between adjacent teeth that are large enough to accommodate all the teeth comfortably in the space-restricted jaw.

Of the many controversies surrounding the interpretation of developmental interactions between body components during ontogenesis, it is important to consider briefly the question of negative developmental interactions in animals. The operation of at least some degree of negative interaction between developing structures has been demonstrated in the growing turtle carapace (Mosiman, 1965), in echinoid plates (Joysey, 1959; Raup, 1968), in mouse vertebrae (Bailey, 1956), and between the developing tibia and fibula of the chicken (Hampé, 1960). Walruses and sabre tooth cats provide examples of phylogenetic increase in canine size at the expense of anterior premolar size (Rensch, 1969). Kurten (1967), however, cautioned against interpreting these phylogenetic events as evidence for negative interaction between canine and premolar tooth size. Instead he claimed that within a population (e.g. of sabre toothed cats), the intertooth size correlation is in fact positive.

Although Sofaer illustrated his model of compensatory tooth size interaction with references to experimental animal studies (Gruneberg, 1951; Grewal, 1962; van Valen, 1962; Gould & Garwood, 1969), none of these studies in fact demonstrated a clearly defined negative interaction pattern across the entire dentition. Gruneberg (1951) considered the possibility that mouse molars might compete for a limited amount of dental lamina: if the first molars were larger than average, the third molars would consequently be smaller than average. Re-examination of Gruneberg's suggestion by Grewal (1962), however, failed to support it. Although a study of the cheek teeth in *Peromyscus leucopus* provided strong evidence for negative interaction between last molars and those mesial to them, surprisingly a positive relation between second and first molar size was noted (van Valen, 1962). Evidence for similar negative interaction between the last molar and its more mesial neighbour was provided in *Nesophontes* and in *Oryzomys*, but again not between the second molar and its mesial neighbour (Gould & Garwood, 1969). That molecular mechanisms involved in cell–cell interactions may set limits on morphological patterns, has been well documented (Seegmiller, Fraser & Clarke, 1977; Odell *et al.*, 1980; Wessels, 1982). The exact nature of mechanochemical interactions governing the extracellular requirements of contact dependence is, however, not well understood.

Clearly there are two points at issue here – does selection favour a harmonious tooth size: jaw size relationship? Is there empirical evidence of compensatory tooth size interactions between early and later developing teeth? If selection is seen as the nonrandom differential perpetuation of varying genomes (Peters, 1983), which results from some form of reproductive advantage conferred by an adaptation, then one would be hard pressed to show how a harmonious reduction of jaw and tooth size could have been selected for, *ab initio*. Burian (1983) in fact warned against the dangers of interpreting morphological features with high engineering fitness as having necessarily been the results of natural selection.

Aside from stressing the unit of the genome during evolutionary change, the holistic view of phenotypic modification as expressed by Lewontin

(1974), Mayr (1976), Riedl (1978), Gould & Lewontin (1979) and Wright (1980), holds that the total genome evolves as a result of selection on the total phenotype, rather than on individual characters or genes. Expanding this view, Riedl (1978) predicted that through selection for interdependence, the pattern of genetic interconnectedness of characters will mimic their functional interdependency. Cheverud (1982a) suggested that elements contained in sets of traits which are highly integrated genetically (G-sets) will show strongly correlated responses to evolutionary forces acting on any individual element within the set. If it can be shown that those sets of traits that demonstrate developmental and functional interdependence (F-sets, Olson & Miller, 1958) also display genetic interdependence, then one may reasonably conclude that all these traits will react to evolutionary forces as a unit.

Functional and developmental relationships within the cranium have been extensively analysed by Moss and his associates (Moss, 1973; Moss & Salentijn, 1969, 1970, 1971; Moss & Young, 1960), who placed dentition into an oral–skeletal subunit of the orofacial matrix. In analysing a number of F- and G-sets within the orofacial matrix of rhesus macaques, Cheverud (1982a) was able to demonstrate a lack of correspondence between genetic and phenotypic patterns of integration, with respect especially to the main orofacial functional component. This observed low correlation between F- and G-sets indicates that units within the orofacial matrix will not necessarily respond to evolutionary forces as a unit, and this inference, in turn, would throw doubt on the validity of Sofaer's model of harmoniously integrated jaw–tooth reduction (only if the jaw–tooth complex comprised F- and G- sets). However, the tight genetic integration demonstrated by Cheverud (1982a) also points away from a mosaic pattern of orofacial evolution in which different units would evolve independently. Harmonious tooth–jaw reduction may of course be considered to be an 'exaptationist' phenomenon (Gould & Vrba, 1982), in other words, a coopted feature whose current advantage is a by-product rather than a result of selection. Of the human condition, one may say that disharmony rather than harmony characterises the tooth–jaw relationships of younger generations. Whether this is due to a relaxation of selection forces, or to a failure of maxillofacial integration, or to other factors remains debatable.

Knowledge of compensatory tooth size interactions in man is based largely on studies by Townsend & Brown (1980), Mizoguchi (1981b, 1983a) and Kieser *et al.* (1985f, 1986c). Most of these, however, do not support Sofaer's hypothesis. Using asymmetry correlations involving adjacent teeth, Townsend & Brown (1980) were unable to confirm the existence of compensatory interactions between adjacent teeth in Australian Aboriginals. In contrast, Mizoguchi (1981b) reported that a proportion of variation within each morphological tooth unit might be ascribed to compensatory growth of later developing teeth. From a more recent study, however, Mizoguchi (1983a) concluded that there was little or no evidence of compensatory growth of later developing teeth within the dental arcade.

Possible compensatory tooth size interactions between early and later developing teeth within the same tooth class have recently been evaluated in the Lengua Indians and in a sample of caucasoid children with 'ideal' occlusions (Kieser *et al.*, 1985f, 1986c). Mean BL and MD diameters for first and second molars, premolars and incisors were calculated, and those teeth that were either larger or smaller than the population mean were identified. To test Sofaer's model – that is, that a negative relation between first developing teeth and their later developing neighbours should be demonstrable – they explored tooth size correlations within the bivariate data matrix, using Pearson's product moment correlation coefficients and chi-square tests. These authors realised, however, that a quantitative analysis of compensatory tooth size interaction would require consideration also of arcade length. Hence, in a series of regression analyses, each MD dimension was regressed on that individual's total arcade length. By using MD dimensions of the left quadrant to obtain an estimate of arcade length when analysing tooth size correlations on the right side, and vice versa, they avoided spurious relationships which could have occurred if dependent variables were treated as components of the independent variable.

Their results revealed that in all cases there existed a highly significant ($p < 0.001$) positive correlation between larger than average (or smaller than average) first teeth and their later developing neighbours. For example, of the first maxillary molars with above average BL dimensions, a full 74.4% were associated with second molars that also had larger BL values than the mean. Only 25.5% of second molars associated with larger than normal first molars had less than average BL diameters. The converse also holds: of the first maxillary molars that had smaller than average BL dimensions, 82.6% also had smaller than average neighbours, but 17.3% were associated with larger than average second molars. The strong positive correlation between first and second teeth within the same tooth class was further illustrated when the actual deviations from the mean for early and later developing neighbours were correlated – Pearsonian correlation coefficients being highly significant in all cases ($p < 0.001$) with the exception of the mandibular incisors of females.

Similar results were obtained for correlations between first and second teeth corrected for arcade length. The authors argued that, if a first tooth was larger than expected for the individual's arcade size, then a smaller than expected second tooth would indicate the presence of tooth size compensation and vice versa. However, the absence of negative correlations prompted these authors to reject Sofaer's hypothesis.

Kieser and his associates (1985f, 1986c) may, however, not have provided unequivocal evidence in rejection of Sofaer's hypothesis. Their results indicated that there was a positive correlation between P1 and P2. Sofaer (1973) had, however, stated that a negative association could be expected between early and later developing teeth, a fact which Kieser *et al.* (1985f, 1986c) had overlooked. The second premolar is in fact the first to develop. Moreover, Harzer (1987) has recently published indirect evidence for the existence of compensatory interactions between adjacent teeth during early odontogenesis.

In a recent investigation Harris & Bailit (1988) examined multivariate correlation matrices in an extremely large sample ($n = 2650$) of individuals from the Solomon Islands. Their findings indicated that individuals with large teeth in one part of the arch tended to have above average tooth sizes elsewhere in the arch. Furthermore, there was a suggestion that cheek teeth were weighted slightly more than anterior teeth. Hence this study both confirms and amplifies earlier observations that suggested that individual teeth within the dental arcade covary positively.

# 9 *Fluctuating odontometric asymmetry*

## 9.1 Introduction

In a thoughtful review of asymmetry in animals, Neville (1976) traced the origins of asymmetry back to a unique quality of the element carbon. With its valency of four, chemical groups may be bonded to carbon in one of two spatial configurations, resulting in the formation of two enantiomorphic isomers, which are mirror images of one another. At a slightly higher level, Neville related asymmetry to the sense of chirality or twist of biological helices. Since all biological helices are $\alpha$-helices, the right-handed twist being favoured by hydrogen bonds, and since these helices are the same on both sides of the body, Neville reasoned that they too represented a type of structural asymmetry present in all living beings.

The origins of interest in the unequal manifestation of bilaterally symmetric structures in man and other animals is probably pre-Aristotelian. Modern interest in asymmetry originated with the writings of Charles Darwin and centred around the question of its inheritance. Darwin noted that even though '. . . it might have been anticipated, that deviations from the law of symmetry would not have been inherited, . . . a man who is left-handed, and a shell in which the spire turns in the wrong direction, are departures from the normal asymmetrical conditions, and they are well known to be inherited' (Darwin, 1905, p. 545). Early observations on fiddler crabs and flat fishes had led Yerkes (1901) and Jordan & Evermann (1896) to assert that even though dextrality or sinistrality may not be inherited, the condition of asymmetry clearly was. Inherited asymmetry continued to be of interest during the first two decades of this century, being the subject of early studies on human fingerprint patterning (Wilder, 1904) and of human handedness (Jordan, 1914). Similarly, Przibram (1907) investigated the patterns of inheritance of dissimilar eye colouring in domestic cats. Accordingly, by the early twenties it had become accepted that asymmetry was an inherited condition. In their prolix review of asymmetry, Sumner & Huestis (1921) are led to conclude: '. . . that all these cases of normal asymmetry are hereditary in a real sense hardly needs argument. The sole alternative explanation, namely, that they are due to environmental influences which affect the two sides unequally . . . needs only to be stated in order to be refuted' (p. 446). Reviewing numerous instances of asymmetry in

96

animals, Ludwig (1932) concluded that bilateral symmetry was never quite perfect in living organisms: the left and the right sides of each individual within each species always differed somewhat. The nagging question remained – how could small, seemingly random left–right differences in animals be accounted for by hereditary factors? This question was to remain begging for over twenty years.

As early as 1942, Waddington suggested that the differentiation of any particular structure will follow a stabilised developmental pathway, even though the underlying genotype may differ somewhat in each case. Describing these stabilised or buffered pathways of histogenesis as creodes, Waddington suggested further that they allowed the developing organism to absorb a certain amount of environmental variation without exhibiting any marked alteration in the phenotypic end-product (Waddington, 1942). Applying Waddington's canalisation theory to observed asymmetry in *Drosophila*, Mather (1953) suggested that, although natural selection appeared to have adjusted the translatory process of the genotype so as to yield a relatively uniform developmental end-product, the course of development may still be upset by outside influences. He concluded that the study of asymmetry therefore affords a means of investigating the genotypic stabilisation of developmental processes. Mather argued further that if the intrinsic developmental paths were the same on both sides of an individual, then the expression of a particular character will be perfectly symmetrical, barring local disturbances. Any asymmetry which may appear will be an expression of such local disturbances, whether arising from differences of environment or errors of cell development; but since such disturbances should have equal chances of affecting each side, the average expression of the character should be symmetrical, even though any given individual may differ bilaterally. Mather therefore identified two very important aspects of this type of asymmetry: it is random in its distribution and it may be employed as a measure of developmental destabilisation. Referring to these local destabilising factors as developmental noise, Waddington (1957) argued further that for an individual to be perfectly symmetrical, it must either possess the capacity to resist the developmental noise (i.e. to buffer against it) or the noise must be insufficient to influence normal morphogenesis.

In a landmark paper published in 1962, van Valen classified asymmetry into three categories: antisymmetry, directional asymmetry and fluctuating asymmetry. Antisymmetry occurs when a difference between sides is regularly induced by competitive interaction between sides, for example in human sinistrality or dextrality. Directional asymmetry occurs whenever there is normally greater development of a structure on one side of the plane of symmetry. Directional asymmetry, van Valen suggested, may best be demonstrated by the systematic difference in mean values of a character between two sides of the body, and is possibly best exemplified by the development of the mammalian heart. Fluctuating asymmetry, on the other hand, reflects the inability of an organism to develop along precisely determined paths on both sides of the body.

## 9.2    Early investigations

Much of the early impetus for the study of fluctuating asymmetry came from a desire to demonstrate its noninheritance (see for example Thoday, 1958; Beardmore, 1960; Reeve, 1960; Mason, Ehrlich & Emmel, 1967). Arguing that inbreeding would lower the developmental buffering capacity of an individual, Niswander & Chung (1965) and Bader (1965) were able to show an increase in the fluctuating asymmetry of dental dimensions in consanguineous Japanese and in inbred laboratory mice. In a significant departure from these genetically orientated investigations, Garn and his co-workers documented in some detail, the empirical manifestations of dental asymmetry in humans (Garn *et al.*, 1965b, 1966, 1967b). They found, for instance, that odontometric asymmetry was randomly distributed with respect to side, and that larger teeth were characterised by absolutely larger asymmetry values. They demonstrated also that asymmetry was more marked in children lacking one or more of their third molars, that asymmetry was greater in the more distal teeth in each morphological class, and that both dimensions (BL and MD) displayed similar levels of asymmetry. A final finding of note was that males showed greater levels of asymmetry than females. Accordingly, they postulated that the paired X chromosome of the female would steer her on a more consistent developmental course, and hence would result in less asymmetry than in males (Garn *et al.*, 1965c). Finally, they stated simply that asymmetry rarely added up to a significant bilateral discrepancy, and hence could not be considered to be of major clinical orthodontic concern. It was probably for this reason that interest in the dental manifestations of asymmetry in humans waned towards the late sixties.

The current resurgence of interest in human dental asymmetry is due largely to the work of Bailit and Siegel. Realising the importance of finding a reliable measure of a population's response to adverse environmental and genetic conditions, Bailit and his associates examined the usefulness of differences in the size of antimeric teeth as measures of fluctuating asymmetry, which in turn, they believed, could be employed as a measure of environmental stress (Bailit *et al.*, 1970). For their study, they chose four widely differing human population samples. The first were the inhabitants of Tristan da Cunha, a small south Atlantic island (Bailit, Damon & Damon, 1966). Largely as a result of its geographic isolation (2898 km from the nearest land mass) and also of the small size of its founding population, the population had one of the highest levels of inbreeding revealed in man. An inhospitable climate and poor terrain contributed further to the high levels of environmental stress to which the Tristanites were exposed. The second population investigated lived in relatively idyllic conditions on the island of Bougainville in the Solomon chain. There was no evidence of inbreeding, and this group was found to be relatively disease-free and well nourished. The third population, although resident also on an island in the Solomon group, lived under very different conditions indeed, on the rugged, humid and wet island of Kwaio which had very little contact with the outside world. Heath facilities were almost nonexistent and malnutrition was rife. Dental asymmetry in these three populations was compared to that found in Boston school children. Arguing that if stress and asymmetry are positively correlated, Bailit and his associates expected the population rankings in order of decreasing asymmetry to be: Tristanites > Kwaio > Nasioi of Bourgainville > Bostonians. By using intraclass correlation coefficients between pairs of antimeric teeth for each individual, Bailit *et al.* were able to show that the trends in mean asymmetry among the populations were as predicted, the Tristanites being the most asymmetric and Bostonians the least (Figure 9.1). Interestingly, the differences in asymmetry levels between the sexes were not statistically significant. Another feature that was not statistically significant was the correlation between the coefficients of inbreeding and the levels of asymmetry; this observation led Bailit *et al.* to conclude that inbreeding was in fact unrelated to dental asymmetry. The inevitable question remained: what was the nature of the 'environmental stress'?

## 9.3 Stress and asymmetry

Major contributions to the modern understanding of stress in general and stress as a causative agent in dental asymmetry were made by Selye and by Siegel and Doyle. The word stress is an aphetism of the old French word *estrecier*, which in its earlier usage referred to forcible constraint or incarceration (Onions, 1973). Largely as a result of Hooke's investigations into the elasticity of springs, stress had, by the late seventeenth century, become an engineering or a physics term for the ratio of an internal force brought about when a substance is distorted, to the area over which that force acts (Hinkle, 1973). By the early twentieth century, however, stress had become also a medical term. Sir William Osler (1910), for instance, in referring to a group of twenty physicians, noted that the '. . . incessant treadmill of the practice of medicine' had placed these individuals under 'stress' which, he argued, might in turn be seen as the basic causative agent of angina pectoris.

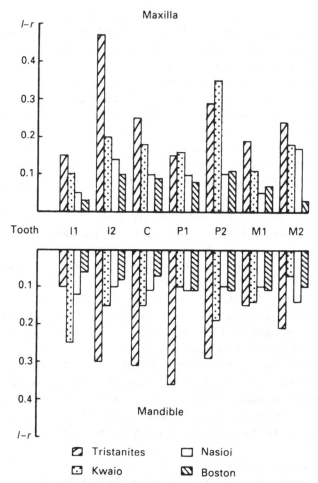

Figure 9.1. Fluctuating odontometric asymmetry as an indicator of stress in four human populations. Asymmetry, $1-r$, is taken as the proportion of the total variance resulting from intra-individual variation between antimeres. (Computed from figures in Bailit *et al.*, 1970.)

One of the first laboratory investigations into stress was carried out some 25 years later by Walter Cannon (1935), who observed the reactions of the sympathetic nervous system and the adrenal medulla in laboratory animals that he had subjected to a variety of adverse conditions such as hypothermia, loss of blood and lack of oxygen. Cannon formally noted that these animals had been 'stressed' and he suggested that the levels of stress might in fact be measurable. By far the most comprehensive series of investigations into the nature and effects of stress have since been conducted by the Canadian investigator Hans Selye. In his description of

the 'stress syndrome' or the 'general adaptation syndrome', Selye documented the central role played by the anterior pituitary and the adrenal cortex
in the body's response to environmental stressors, and underlined the
nonspecificity of these stressors or damaging agents (Selye, 1952, 1973).

During the seventies, Siegel and Doyle systematically investigated the
asymmetrogenic effects of various stressors on experimental animals at the
Environmental Studies Laboratory in Pittsburgh. They were also to
demonstrate that heat, cold and audiogenic stress increased the fluctuating
asymmetry in the teeth and limbs of rodents (Siegel & Smookler, 1973;
Siegel & Doyle, 1975a, b; Siegel, Doyle & Kelley, 1977; Doyle, Kelley &
Siegel, 1977; Sciulli *et al.*, 1979). Fluctuating asymmetry, therefore, became
known as a host response to stress, the term stress being interpreted broadly
as defined by Selye's General Adaptation Syndrome.

## 9.4 Human odontometric asymmetry

The return of interest in human dental asymmetry was heralded by
a paper by Suarez in which he used levels of asymmetry to test Brace's
probable mutation effect hypothesis (discussed in Chapter 5). Suarez (1974)
based his argument on two assumptions. The first assumption was that
stabilising selection would operate equally on the left and right sides of a
dental arch, the second that the correlation of teeth on the left side of the jaw
and their respective homologues on the right side of the jaw would be the
same for different human populations. Attempting to test the validity of
applying the probable mutation effect to the observed trend towards
human tooth size reduction, Suarez calculated the variability of antimeric
dental dimensions in samples of neanderthaloids and of modern Ohio
caucasoids. He expected the Neanderthal population to show lower levels
of asymmetry than modern man. His findings, however, were the direct
opposite: Neanderthal dental asymmetry was significantly greater than
that of Ohio caucasoids. From these results, Suarez drew two conclusions.
First, the probable mutation effect failed to explain human dental size
reduction and secondly, drawing on the results obtained on wild and inbred
lines of mice by Bader (1965), that Neanderthal populations were more
inbred than those of modern man.

Doyle & Johnston (1977) disputed these conclusions of Suarez, suggesting rather that environmental stress was the asymmetrogenic agent. They
cited Mayr's thesis that the adjustment of a species to extended periods of
environmental stress is brought about through evolution which in turn
serves to readjust the species to a changed environment (Mayr, 1966).
Doyle and Johnston argued that the primary mechanism of adaptation to
environmental stress employed by hominids was cultural. Therefore, they
argued, the degree of stress experienced by an individual within a

population is inversely related to the available degree of technological control over the environment. By demonstrating levels of fluctuating odontometric asymmetry in Pueblo Indians and in Eskimoes that were comparable to those found in Neanderthals, Doyle and Johnston were able to reject the inbreeding theory of Suarez.

Continued interest in fluctuating dental asymmetry in humans has since centred around two broad fields of enquiry: genetic and ethnogeographic.

## 9.5    Genetic investigations

Evidence for a genetic basis of asymmetry in the dental dimensions of humans was first sought by Potter & Nance (1976). They surveyed 75 pairs of twins drawn from the Indiana University Twin Panel. By contrasting the levels of asymmetry between the 43 pairs of monozygotic and 32 pairs of dizygotic twins, Potter and Nance were able to show that there was little or no evidence for a genetic basis of fluctuating odontometric asymmetry (Figure 9.2).

On a slightly different level stand those studies of fluctuating dental asymmetry in Down's Syndrome (trisomy 21). Citing an earlier study by Adams & Niswander (1967), Barden (1980) argued that increased levels of fluctuating asymmetry would be indicative of a loss of genetic control over development. Because previous investigations had shown the major effect of the trisomic condition in Down's Syndrome to be a decrease in overall developmental stability of the individual (Shapiro, 1971, 1975), Barden predicted and showed elevated levels of dental size asymmetry in Down's Syndrome subjects. He found a disproportionately high increase in asymmetry in those teeth that have traditionally been accepted as being the least stable. These findings echoed those of Garn *et al.* (1970) and were confirmed in a similar survey in Australia (Townsend, 1983b).

Another investigation relating a genetic malformation with high levels of developmental instability involved individuals with cleft lip, with or without cleft palate [CL(P)]. In populations of patients with CL(P) a distinction needs to be drawn between those individuals in whom the cleft is sporadic and those who have a familial history of CL(P). The latter group, in turn, may be subdivided into individuals who have clefts associated with other malformations and those who do not. It is interesting that investigations into familial CL(P) without associated malformations have found elevated levels of asymmetry in structures remote from the site of the cleft. Similarly elevated asymmetry levels were recorded in their noncleft relatives, but not in sporadic CL(P) cases or their relatives (Adams & Niswander, 1967; Woolf & Gianas, 1976, 1977). One possible explanation is that those genes common to familial CL(P) and their unaffected relatives might be responsible for the high levels of developmental asymmetry.

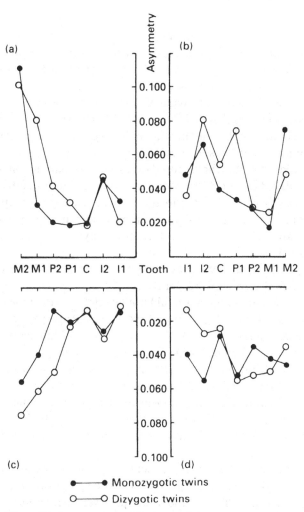

Figure 9.2. A comparison of the levels of dental asymmetry in di- and monozygotic twins: (a) maxillary teeth, MD dimension, (b) BL dimension, (c) mandibular teeth, MD dimension, (d) BL dimension. (Computed from figures in Potter & Nance, 1976.)

Realising this, Sofaer (1979) conducted an investigation into fluctuating asymmetry of dental dimensions in a group of 77 patients with CL(P). By dividing his population into a familial and a sporadic group, Sofaer hoped to demonstrate lower levels of dental asymmetry in the sporadic group who, he surmised, would lack the inherited instability of development which characterised the familial CL(P) group. Although Sofaer's results pointed to an increase in dental asymmetry in CL(P) patients, his comparisons between familial and sporadic groups were inconclusive.

## 9.6     Ethnogeographic investigations

Application of the principles of Waddingtonian canalisation to human odontogenesis permits the analysis of fluctuating odontometric asymmetry in terms of ethnogeographic differences. One of the first investigations in which dental asymmetry was contrasted among human skeletal populations was that of Perzigian (1977). He estimated asymmetry in prehistoric hunters (from the Indian Knoll site), later aboriginal farming groups (Campbell and Larson sites) and a modern cadaver population (Hamann-Todd). Of the three populations, the Indian Knoll hunter-gatherers displayed the greatest asymmetry, with no significant differences between the levels of asymmetry in the Campbell/Larson group or the Hamman collection. These results prompted Perzigian to conclude that environmentally mediated growth disturbances may be sensitively reflected by dental asymmetry.

Di Bennardo & Bailit (1978) investigated, in some detail, the possible effects of prenatal stress in the generation of dental asymmetry. They concluded that dental asymmetry could not be related to their measures of prenatal stress (birth weight, gestational age, maternal age, parity, sex, year of birth, socioeconomic status and the coefficient of inbreeding). This conclusion was based upon MD diameters of only upper and lower permanent lateral incisors and first molars, as well as a modified buccolingual diameter of the first molar. It might well be argued that consideration of all the teeth, and a measure of asymmetry which records not merely the absolute deviation from the norm (as would the Naperian log transformed left and right measurements) might have led to a different conclusion.

Further criticism of Perzigian's conclusions was provided by Black (1980) who studied dental asymmetry in 208 Michigan caucasoids. Using antimeric correlations Black found that the level of asymmetry in his population in fact exceeded those of Perzigian's skeletal groups, it being implied that either Perzigian's thesis should be rejected or that their measure of asymmetry was suspect. In the same year a novel symmetry measure was introduced by Harris & Nweeia (1980a), which they felt would circumvent the inefficiency of the more traditional root mean square (RMS) or intraclass correlation coefficients. Harris and Nweeia employed a rescaled antimeric asymmetry measure ($d^*$) in which the side difference $(l-r)$ was divided by the mean size of the left and right teeth, an absolute value taken to preclude positive and negative values cancelling one another. Its obvious advantage lies in the fact that it eliminates size differences among teeth and so avoids the difficulty that a given amount of asymmetry in a small sized tooth pair will have greater biological significance than the same amount in a large sized pair.

The application of rescaled values to a sample of 57 Ticuna Indians of Colombia and to previously published data for Ohio caucasoids, Pueblan Indians and Eskimos led Harris and Nweeia to conclude that the Ticuna showed odontometric asymmetry levels greater than those of Ohio caucasoids, but significantly lower than those of Eskimos or Pueblans. The males were more asymmetric than the females, and also maxillary teeth were more asymmetric than mandibular teeth. In a similar investigation into a sample of 392 Australian Aboriginals, males were again shown to be more asymmetric than females (Townsend & Brown, 1980), these results providing support for Garn's theory that the paired X chromosome in females exerted greater control over odontogenesis (Garn *et al.*, 1966). Townsend and Brown postulated that the longer period of tooth formation in males would provide greater opportunity for environmental disturbances to disrupt tooth morphogenesis. Interestingly, when Townsend & Garcia-Godoy (1984) investigated asymmetry in the deciduous teeth of Dominican mulatto children, they were unable to demonstrate sexual dimorphism in the levels of asymmetry.

Garn and his co-workers called into question the use of population differences in the magnitudes of dental asymmetry. They have argued that the entire population range of asymmetry values can be explained on the basis of sample size, the largest and the smallest values for asymmetry being found in the smallest sample sizes (Garn, Smith & Moyers, 1981, p. 33). Further examples of this argument centre about the claim that, when simple correlation coefficients or variance measures are used, sample sizes of several hundred are needed to detect population differences in dental asymmetry (Garn, Cole & Smith, 1979; Smith, Garn & Cole, 1982). Inspection of the relationship between sample size and the RMS asymmetry that Smith and her co-authors examined, using computer generated independent random subsamples from a population of more than 3000 individuals (Figure 9.3), reveals that the lowest and highest estimates of RMS are associated with sample sizes of less than 75. However, an increase in sample size up to 400 results in a gain in accuracy of only 0.04 mm. This suggests that sample sizes in excess of 75 may confidently be used for the computation of RMS asymmetry values. Unfortunately, Garn and his associates did not consider the newly introduced rescaled asymmetry value of Harris & Nweeia (1980a), neither did Greene (1984) in his evaluation of asymmetry in a Nubian skeletal population.

In their recent evaluation of fluctuating odontometric asymmetry as an indicator of environmental stress, Kieser *et al.* (1986a) used the rescaled asymmetry measure on dental casts of 202 living Lengua Indians from the Chaco area of Paraguay, and of a caucasoid sample (Kieser & Groeneveld, 1986a). They reasoned, however, that the rescaled measure views each tooth pair as an independent variable, which it clearly is not. They therefore introduced the more sophisticated Euclidean map analysis, in which each mouth was viewed as an independent variable consisting of fourteen associated antimeric pairs (Kieser *et al.*, 1986b). Intraindividual right–left size differences between antimeres for all teeth save M3, in both dimensions, were plotted on two Euclidean axes. Each Euclidean distance from the origin (calculated as $\sqrt{[(MD)^2+(BL)^2]}$) thus represents an antimeric asymmetry value. When taken collectively such values form a Euclidean map that charts the full mouth asymmetry pattern of that individual. Tight point clustering about the origin indicates low asymmetry levels, while scattering indicates high degrees of asymmetry and hence relatively low odontometric canalisation.

The sum of the Euclidean distances for all antimeres per person, divided by the mean individual tooth size and by the number of tooth pairs considered, provides a single measure of individual odontometric asymmetry corrected for tooth size differences. The relationship may be expressed as follows:

$$\text{Euclidean asymmetry} = \frac{\Sigma\ \text{distance}}{n(\bar{x}\text{Ts})}$$

Figure 9.3. The relationship between sample size and the root mean square asymmetry measure, RMS. Sample sizes of less than 75 are associated with the highest levels of asymmetry, whilst an increase from 75 to 400 results in a gain in accuracy of around 0.04 mm. (Adapted from Smith *et al.*, 1982.)

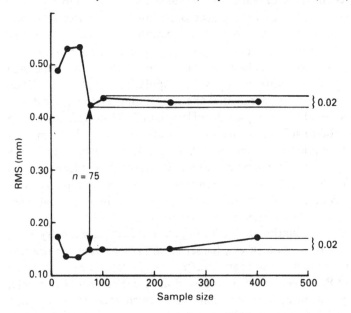

Table 9.1. *Principal causes of childhood mortality in rural Paraguay and in South African caucasoids (Grassi & Salinas, 1979; Stein, 1981)*

| Rural Paraguay | Caucasoid |
| --- | --- |
| 1. Enteric disease | 1. Accidents |
| 2. Pneumonia | 2. Cot deaths |
| 3. Measles | 3. Congenital malformations |
| 4. Deficiency diseases | 4. Baby battering |

where $\bar{x}Ts$ = mean individual tooth size for both dimensions and n = number of tooth pairs.

As expected, the Lengua Indians demonstrated similar levels of asymmetry to the Ticuna Indians of Colombia, but were significantly more asymmetric than contemporary caucasoids. The heavy burden of malnutrition and illness of rural Paraguayan populations is well documented. Children in their tooth-forming years are the most vulnerable (Canese, 1978; Canese & Brice, 1978; Grassi & Salinas, 1979). The relatively low disease burden of the caucasoid sample becomes immediately obvious when the four principal causes of mortality in childhood of the two populations are contrasted (Table 9.1).

Another finding of note was that older, less acculturated Lengua Indians were significantly more asymmetric than younger individuals who had been increasingly exposed to western foodstuffs and medicine. Interestingly, analyses of variance confirmed a lack of sexual dimorphism in asymmetry levels for both populations.

Addressing the problems of small sample size and the inherent incompleteness of fossil data in the evaluation of metric asymmetry, Kieser & Groeneveld (1986b) introduced the application of cluster sampling techniques together with a modified factorial analysis of variance. When metric asymmetry in *Australopithecus*, *Homo erectus* and *Homo neanderthalensis* was calculated, MD diameters emerged as the most asymmetric maxillary teeth exhibiting significantly greater degrees of asymmetry than mandibular teeth. Neanderthal tooth measurements were found to be significantly more asymmetric than those of either australopithecines or *Homo erectus*.

The usefulness of asymmetry studies in the field of community medicine has recently been demonstrated by an investigation which contrasted asymmetry levels in South African Negroes and caucasoids (Kieser & Groeneveld, 1988c). The disease and malnutrition burdens of South African

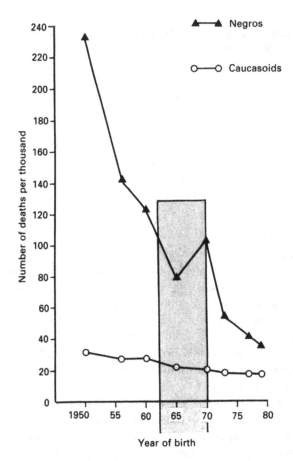

Figure 9.4. Infant mortality rates for urban South African blacks and caucasoids. (Computed from data in Stein, 1981.) Shaded area represents range of individuals considered in the present analysis. (Kieser & Groeneveld, 1988c.)

Negroes are well documented (see, for example, Unterhalter, 1982; Irwig & Ingle, 1984; Botha, Bradshaw & Gonin, 1985). Infant mortality rates are high compared to caucasoids (Figure 9.4), as are the levels of morbidity during the tooth-forming years. In common with other westernised communities, the principal causes of death in childhood for South African caucasoids are non-infective (being accidents, cot deaths and malignancies). In contrast, the causes of death in childhood for Negroes are infections (e.g. gastroenteritis, pneumonia and tuberculosis) and malnutrition (Stein, 1981). The nutritional inadequacies of the partially westernised diet of urban Negroes have been emphasised by numerous authors. With the transition from rural to urban living have come recognisable deficiencies in riboflavin, niacin, tryptophan, protein, calcium, folic acid, dietary fibre and iron (see, for example, Walker, 1982; Steyn *et al.*, 1986).

It therefore comes as no surprise that the levels of odontometric asymmetry in South African Negroes exceed those of caucasoids (Kieser & Groeneveld, 1988c). What is surprising, however, is that these levels also exceed those reported for the Lengua Indians of Paraguay – a population with an apparently similar illness and malnutrition pattern. These authors concluded that the disproportionally high values of asymmetry recorded in South African Negroes probably resulted from the combination of a high disease and malnutrition burden and from a decreased individual buffering ability. However, there may be factors other than such stress which are responsible for the elevated levels of asymmetry in Negroes.

In a recent study, Boklage (1987) criticised the inference of a causal link between stress and asymmetry. Boklage (1987) argued that because twinning in humans had been repeatedly associated with excesses of prenatal, perinatal and neonatal anomalies, twins could be considered to be exposed to greater environmental stress than singletons. Furthermore, because monozygous twinning was associated with a greater degree of embryonic disturbance than dizygous twinning, one might expect fluctuating asymmetry in monozygous twins to exceed that found in dizygous twins, which would in turn be higher again than those of singletons. Boklage's analysis of 84 monozygous and 64 dizygous twins revealed the opposite: twinning of either zygosity was associated with a reduction, not an elevation of fluctuating asymmetry. Clearly, either the underlying assumption of increased stress with twinning is erroneous, or the premise that stress and asymmetry are associated is questionable. Boklage opted for the second, but failed to provide a coherent explanation.

Evidence against Boklage's suggestion has recently come from Mizoguchi (1987) and from Sharma & Corruccini (1987). In both these investigations the genetic–environmental variance for fluctuating asymmetry was partitioned, using mono- and dizygotic twin pairs. Both studies confirmed the environmental influence on dental asymmetry. Sharma & Corruccini (1987) concluded further that the environmental influence frequently differed by zygosity and often mimicked the genetic variance.

Developmental stability may be evaluated from phenotypic variability (Chapter 7) as measured by the coefficient of variation (CV). Leary, Allendorf & Knudsen (1985) demonstrated a positive correlation between the CV and the levels of asymmetry in specific traits of the rainbow trout (*Salmo gairdneri*). However, these authors felt that a high CV might indicate either a high degree of developmental instability or a large amount of genetic variability which has been expressed accurately in the phenotype. Hence, they reasoned, asymmetry could be considered to be a better

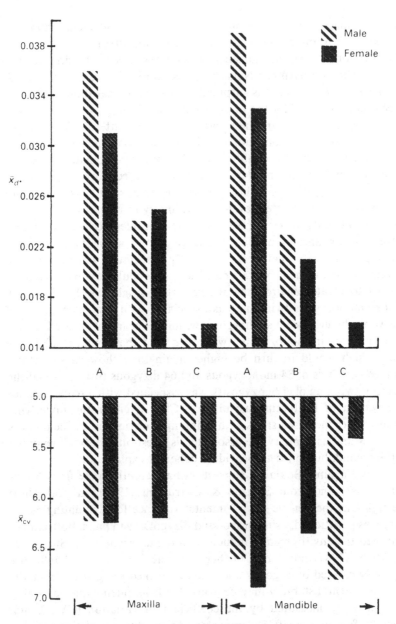

Figure 9.5. Mean rescaled asymmetry ($\bar{x}_{d*}$) and mean coefficients of variation ($\bar{x}_{CV}$) maxillary and mandibular teeth in South African Negroes (a) and caucasoids (c) and in Paraguayan Lengua Indians (b).

measure of developmental stability than the CV. The relationship between fluctuating human odontometric asymmetry and meristic variability (CV) has recently come under scrutiny. Kieser & Groeneveld (1988d) compared CV's and asymmetry values in Lengua Indians, caucasoids and Negroes. Significant levels of Kendall's tau suggested a high concordance between asymmetry and metric variability in the maxillary teeth of all three populations (Figure 9.5). This pattern, however, was not as strongly expressed in the mandibular teeth, where only male Lengua Indians and female caucasoids displayed significant levels of association. Unlike the CV, fluctuating asymmetry is related purely to developmental noise and has almost no additive genetic component. The CV, on the other hand, relates to both genetic and environmental variability. Thus, if the dimensions of lower and upper teeth are largely independently determined, then differences in their CV's may merely reflect differences in the amounts of genetic variability expressed in their phenotype. It was concluded that these results provide further support for the suggestion that asymmetry is a better estimator of developmental instability than of dimensional variability as measured by the CV.

In conclusion, the biological basis of the observed relationship between fluctuating asymmetry and developmental instability requires further investigation. Of particular interest would be further population studies, investigations into possible specific asymmetrogenic agents in man (e.g. nicotine, alcohol, steroids) and, finally, the effects of gestational stressors on dental asymmetry. Nevertheless, the results of the studies presented in this chapter indicate that fluctuating odontometric asymmetry is a valuable tool which may be usefully employed in the fields of epidemiology, embryology, evolution and genetics.

# 10 *Allometric relations of tooth size*

## 10.1 Introduction

Biological scaling is concerned with the effects and consequences of changes in size (Schmidt-Nielsen, 1975). The antiquity of interest in scaling is reflected in its derivation from the Latin word *scala* (staircase), which had by 1662 acquired the meaning of a systematic classification of objects proportionate to one another (Onions, 1973). Twenty-five years earlier, Galileo had published the *Dialogues* in which he discussed the proportional sizes of the bones of different sized animals. Realising that the mass of an animal would increase by the third power of its linear dimensions, Galileo noted that in order to support its body mass, a large animal would need bones that were not only absolutely, but also proportionately larger than those of smaller animals. Large and small animals are thus not geometrically similar; rather their proportions are scaled, such nongeometric scaling being referred to as allometric scaling (from the Greek, *alloios* which means different).

Investigations into the allometry of the human dentition probe the differences in the sizes of teeth or groups of teeth, and then associate these differences with the size or mass of other parts of the body. Allometric studies thus allow one to separate tooth size changes that occur in conjunction with changes in body size from those that occur independently of body size change. Thus, in its broadest sense, allometry allows one to determine whether teeth are larger '. . . all the better to eat you', or whether they are larger merely as a consequence of the enlargement of another part of the body. In summary then, allometry may be defined as '. . . the study of proportion changes correlated with variation in size of either the total organism or the part under consideration' (Gould, 1966, p. 629).

Modern allometric analyses owe much to the seminal work of D'Arcy Thompson, Huxley and Teissier. It was Julian Huxley who showed that a simple power function $Y = bX^a$ might be used to describe correlated growth changes between two parts ($Y$ and $X$) of an organism (Huxley, 1932). Although Huxley's work was confined to an analysis of size-correlated shape changes during growth, his power function has since been applied to an ever increasing spectrum of allometric investigations. Before considering

112

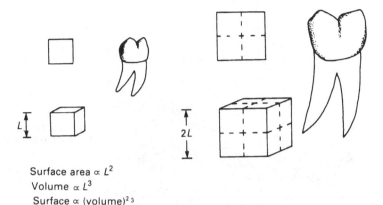

Surface area ∝ $L^2$
Volume ∝ $L^3$
Surface ∝ (volume)$^{2/3}$

Figure 10.1. Changes in mensural elements associated with an increase in tooth size. Under the constraints of isometry, doubling of the length ($L$) of a plane figure such as a square will be accompanied by doubling of the lengths of the other sides, while the surface area will increase as doubling raised to the power two. In a cube, doubling of the side length ($L$) results in an increase in surface area of $(2L)^2$ and an increase in volume of $(2L)^3$. (Adapted from Gingerich & Smith, 1985.)

these, however, it is necessary to consider some of the simple properties of mensuration.

## 10.2    The power formula

Mensuration is the measurement of geometrical quantities, such as the lengths of lines, the size of angles, the areas of surfaces and the volumes of solids. These measures may of course be used to describe the size and shape of an object. It is obvious that a change in any one of these mensural elements will be associated with a change of size and of shape. However, an object may change in size whilst it retains its shape. This is referred to as an isometric change and it involves a constant proportional change of the mensural elements. Consider for instance a plane figure such as a square (Figure 10.1). If the length of one side ($L$) is doubled, then isometric shape change will involve doubling of the length of the other sides. The surface area, however, would increase as doubling raised to the power two. If one now considers a cube, isometric transformation by doubling $L$ will result in a cube of side length $2L$. However, the surface areas of the two cubes will not change in the same ratio as their linear dimensions, rather the surface area increases as $(2L)$ squared or $4L^2$. The surface area of the larger cube will thus be four times that of the smaller cube. Its volume on the other hand will increase by doubling of $L$ cubed, or $(2L)^3$, which is $8L^3$. Hence the volume of the larger cube will be eight times that of the smaller cube. Of great significance is the conclusion that under the constraints of isometry, the

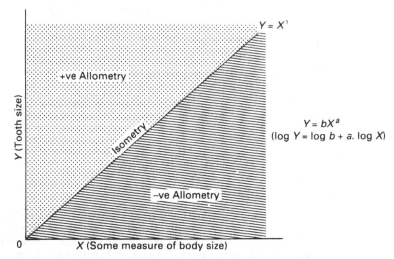

Figure 10.2. Allometric and isometric change in tooth size ($Y$) as a function of some measure of body size ($X$), where $b$ is a constant and $a$ is the allometric coefficient.

surface area of an object increases less rapidly than its volume, in fact it increases in proportion to the power two-thirds of the volume (Gingerich & Smith, 1985; Schmidt-Nielsen, 1984).

Julian Huxley's empirical allometric formula $Y = bX^a$ becomes linear in its logarithmic form $Y = a \log X + \log b$. The advantage of the logarithmic form is that, when the two variables ($Y$ and $X$) are plotted on logarithmic coordinates, the result is a straight line with slope $a$ (the allometric exponent – Gould, 1966).

In its simplest form, then, an allometric analysis involves merely the bivariate regression of a dependent variable (e.g. tooth size) on the ordinate and some estimate of size – the independent variable (e.g. jaw size) – on the abscissa (Fleagle, 1985). Figure 10.2 demonstrates allometric and isometric change in tooth size ($Y$) as a function of some measure of body size ($X$), where $b$ is a constant and $a$ is the allometric coefficient. Here, $Y = X^1$ is the isometric baseline for comparison of the measures of tooth size and body size, where both have the same dimensionality. If the value of $a$ is one, then the two variables scale isometrically. Allometric scaling, however, is indicated by slopes of the regression line that are either larger than one (positive (+ve) allometry) or smaller than one (negative (−ve) allometry). The regression lines for different exponents of the power formula $Y = bX^a$ are given in Figure 10.3. If the dependent variable (tooth size) increases in proportion to the measure of body size, then $a = 1$. If, however, tooth size increases proportionally more rapidly than body size (b), then the slope

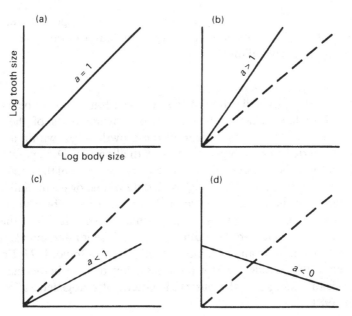

Figure 10.3. Isometric and allometric changes in tooth size as a function of body size. The isometric baseline of slope 1 is shown in (a). When the slope of the regression line is greater than 1, tooth size scales positively allometric to body size (b). A slope of less than 1 indicates negative allometry (c), and a slope of less than zero reflects a negative correlation (d).

will be greater than one (i.e. $a > 1$). A regression line of slope less than one, on the other hand, would indicate that, although tooth size increases with body size, it does so less than proportionally to body size (c). When the slope is less than zero, of course, the dependent variable decreases with increasing body size (d).

Schmidt-Nielsen (1984) pointed out that the power formula $Y = bX^a$ (or its corresponding regression line) has two important numerical terms: the allometric exponent (which is the regression slope – $a$) and the proportionality coefficient $b$ (which is the intercept at unity). The exponent yields information about the proportionality of the changes between the two variables, that is, how they scale together. The proportionality coefficient, on the other hand, provides information about how large the dependent variable is in relation to the independent variable.

## 10.3 Types of allometric analysis

In its earliest application, the term allometry referred only to differential growth, or size-correlated changes fitted by power functions (Hersch, 1955). However, the term has gradually acquired a wider connotation. Today it is applied to three major types of analyses: ontogenetic or developmental allometry, static adult intraspecific allometry and interspecific allometry.

### 10.3.1    Ontogenetic allometry

Huxley originally showed that a power function may be used to describe correlated changes between two growing parts of the body (Huxley, 1932). Such a comparison of the growth stages within a single species is referred to as ontogenetic or growth allometry. Data points used in such investigations may be drawn from a cross-section of the population under consideration, or from the growth data of one or more individuals over a period of time (longitudinal data). It should of course be remembered that the power formula is merely an approximation, since it reflects the sum of composite characters that not only grow at different rates, but also have different times of onset and of completion of growth (Gould, 1977; Ford & Corruccini, 1985; Lande, 1985). Alterations in any of these factors may lead to significant changes in the observed pattern of ontogenetic allometry (Aiello, 1981).

### 10.3.2    Intraspecific (static) allometry

Here the phenotypic endpoints of ontogeny are of interest, and different sized adults of a species are thus compared. If it is assumed that members of the same species have the same adaptive needs, then it may be argued that adult intraspecific allometry represents a true measure of size-related morphological changes that are independent of adaptation (Fleagle, 1985; see also Gould, 1978). It has been suggested also that intraspecific allometry may be useful in the inference of ontogenetic allometry (Zihlman & Cramer, 1978; Corruccini & McHenry, 1979; McHenry & Corruccini, 1981). Shea (1981), however, documented significant differences between intraspecific and ontogenetic scaling in hominids. Others argued that, despite their frequent similarity, there are major theoretical and empirical differences between ontogenetic and intraspecific allometry (Cock, 1966; Cheverud, 1982b).

### 10.3.3    Interspecific allometry

Scaling may be examined also across a range of species that may differ in body size and the samples of which are all at a similar growth stage. This is referred to as interspecific allometry.

As Wolpoff (1985) emphasised, the key to understanding the true nature of allometry is a careful consideration of the differences between these three types of scaling. Originally, Huxley (1932) distinguished between only those allometric trends that accompany the growth of an individual (heterauxesis) and those that emerge from a comparison of many individuals at a single growth stage (allomorphosis). Whilst it has since been shown that scaling at a fixed growth stage (usually at the adult stage) is not the same as inter- and intraspecific, the latter terms are constantly being confused with one another (for a detailed discussion, see Wolpoff, 1985). This confusion arises not so much from the methodology applied to the allometric analyses as from phylogenetic and functional interpretations attached to their results.

Here one might consider the linear relations obtained from intra- and interspecific allometric analyses. When the data are drawn from a single population, the aim of the linear model is to quantify size-related variation within the population (Figure 10.4a). If, on the other hand, a number of different species are being investigated, the intention is to quantify between-population variation, subject to within-population variation (Figure 10.4b). Rayner (1985) stated that in an intraspecific investigation, any variation – after allowance for measurement errors – must be considered as functional. When multiple populations are considered, however, the situation becomes a lot more complicated. The apparent linear relation in Figure 10.4b might suggest that the species investigated vary functionally and may thus be considered to fall on an allometric trajectory, from which deviant species might be isolated. Hence the allometric analysis would allow one to establish a criterion of subtraction, which controls for size-related shape changes, and which facilitates the evaluation of nonallometric adaptive changes (Corruccini, 1972; Pilbeam & Gould, 1974; Gould, 1975; Pirie, 1978; Susman & Creel, 1979; Smith, 1981a; Gingerich & Smith, 1985; Steudel, 1985). However, Shea (1983b, p. 33) cautioned that '. . . this is not simply to assert that one removes size in order to reveal adaptation'. Rayner (1985) argued against preclassification of the species by use of the same functional (or adaptive) criteria as those being tested for in the allometric analysis. If the similarities between the species being compared are due to their common phylogeny and to the constraints of their ontogeny, then these features should not be used also to define the existence of a functional relation beyond the shared features that define the group. Accordingly, Rayner (1985) noted that when one considers the biomechanical design of animals, '. . . there is frequently a concept of pattern such as isometry . . . which describes the "average" shape at any particular size, and mechanical arguments are often phrased in such a way that this single hypothesis appears responsible for all functional variation within the sample' (p. 436). Clearly, adaptation could hardly be viewed as simply as this. It is suggested instead that allometric relationships should be used to predict the constraints on adaptation. 'It is rarely possible to suggest that an organism should be of particular shape or should behave in a particular way for a

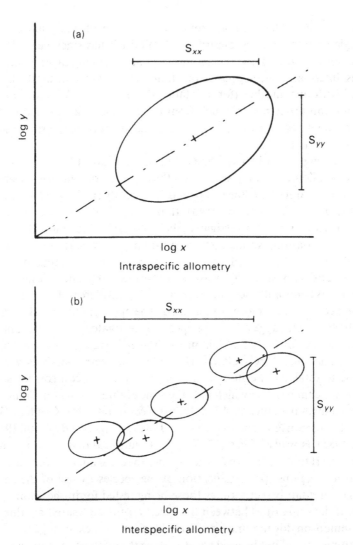

Figure 10.4. Linear relations obtained from intraspecific (a) and interspecific (b) allometric analyses, drawn with similar correlations and data variances ($S_{xx}$, $S_{yy}$). (Adapted from Rayner, 1985.)

single reason, but it is certainly possible to argue that its shape . . . is confined by size-dependent limiting constraints' (Rayner, 1985, p. 437 – see also Clutton-Brock & Harvey, 1979; Alexander, 1982).

## 10.4    Human odontometric allometry

Since its first formulation in 1932, Huxley's power formula has been applied to an ever-increasing spectrum of allometry analyses (Schmidt-Nielsen, 1984). Its application to the dentition of nonhuman primates has centred around the investigation of relationships between tooth size and body size. Such studies, in turn, have proven invaluable in the elucidation of metabolic scaling (Pilbeam & Gould, 1974; Kay, 1975a; Goldstein, Post & Melnick, 1978; Pirie, 1979; Post, Goldstein & Melnick, 1978; Gingerich & Smith, 1985), the investigation of dental sexual dimorphism (Leutenegger & Kelly, 1977; Harvey *et al.*, 1978; Clutton-Brock, 1985; Leutenegger & Cheverud, 1985), and craniodental ontogenetic scaling (Freedman, 1962; Shea, 1983b, 1985a, b). Allometric investigations of dental size correlates are also useful in the prediction of body size from tooth size in fossils (Gingerich, 1977; Kay & Simmons, 1980; Gingerich, Smith & Rosenberg, 1982; Smith, 1985; Steudel, 1985) and in hominid taxonomy (Pilbeam & Gould, 1974; Kay, 1975b; Wolpoff, 1978a, b; Wood, 1979; Wood & Stack, 1980). In sharp contrast to this large body of literature stand the only analyses in which Huxley's power formula has been used on human odontometric data: namely those of Perzigian (1981); Wolpoff (1985); Kieser & Groeneveld (1988g) and of Shea & Gomez (1988).

Perzigian evaluated allometric relationships between tooth size and body size in 156 preagricultural, prehistoric Kentucky Indians who lived between 4000 and 2500 BC. By using femur length as a measure of body size, Perzigian (1981) was able to demonstrate a negative tooth size to body size association. That is, he was able to show that taller individuals had absolutely, but not relatively, larger teeth than shorter individuals. Perzigian's findings are especially enlightening when viewed against more traditional linear correlations. Filipson & Goldson (1963), for instance, reported no significant correlation between stature and the widths of canines and incisors in 110 male army conscripts. Five years later, Garn and his associates examined the linear relationships between tooth size and body size in 109 caucasoid participants in the Fels Longitudinal Study. By correlating standing height at 17 years of age to the MD and BL diameters of all the teeth, Garn *et al.* (1968a) concluded that there existed a low-order relation between tooth size and stature. In 1976 Henderson & Corruccini published the results of an investigation into linear tooth size to body size correlates of 29 to 104 American Negroes. Again the conclusion was reached that the correlation was low, and hence that tooth size to body size inferences in fossil hominids should be considered to be unwarranted (Henderson & Corruccini, 1976). Lavelle (1977) examined linear correlates between molar areas and long bone size in 40 male Anglo-Saxons and also in 40 male nineteenth-century Londoners. Significant correlations are found to centre around the maxillary molars. There followed the work of Anderson, Thompson & Popovich (1977) who sampled 118 males and 102 females at the age of 16 years, from the Burlington Growth Centre, and who were unable to demonstrate significant linear correlations between anterior tooth size and body height. Interestingly, anterior mandibular tooth widths were found to be significantly correlated to body weight, but only in males.

Whilst the studies of Filipson & Goldson (1963), Garn *et al.* (1968a), Henderson & Corruccini (1976), Lavelle (1977) and Anderson *et al.* (1977) were confined to a search for patterns of linear correlation between tooth size and body size, Wolpoff (1985) turned to Huxley's power formula in order to investigate scaling relations of tooth size to body weight and height in Australian Aboriginals. Noting that female tooth size correlations with stature were similar to those of males, but correlations with weight were significantly less, Wolpoff asserted that one might expect females to have relatively larger teeth than do males, as an evolutionary consequence of the elevated masticatory demands placed on them. Interestingly, when Wolpoff determined postcanine areas as fractions of body weight, the results seemed to support his thesis: the relative maxillary postcanine areas, for instance, were found to be 17% greater in females than in males. From the foregoing, it was concluded that females had experienced relatively greater uncoupling of tooth size to body size relations as a response to selective forces that tended to favour relatively larger postcanine areas and narrower ranges of size variation.

One of the main concerns in our attempt to interpret Wolpoff's results is the extent to which variability of the independent variable may have led to an overestimation of the allometric coefficient. If, for instance, it could be shown that the body weight of Australian Aboriginal females had a greater range than that of males, then one might reasonably conclude that the slope of the least squares regression reflects not only the organic correlation between tooth size and body weight, but also the overall variability of body weight in females. That the choice of an independent variable may affect the estimation of scaling patterns had, in fact, been pointed out four years earlier by Smith (1981b). He demonstrated that when the body weight of primates was used as an independent variable, then the scaling with tooth size tended to be negative. When, on the other hand, Smith used cranial length as an independent variable, scaling with tooth size became positive.

It was for this reason that Kieser & Groeneveld (1988g) decided to search for less ambiguous independent variables on which to base their allometric analyses. Centring on four viscerocranial measures (bimaxillary breadth, maxilloalveolar length, maximum projective length of the mandible and bigonial breadth) they were able to explore the scaling of teeth and jaws in a sample of 100 Negro crania of known age and sex.

## 10.4.1 Craniodental allometry

Few recent investigations relate to the scaling of canine size in humans. In a pooled sample of early Kentucky Indians, Perzigian (1981) noted a negative allometry between canine size and femoral length. Similarly, Wolpoff (1985) noted negative scaling of canines to stature in Australian Aboriginals. In contrast Kieser & Groeneveld's (1988g) results indicated a strong positive allometry between canine base area and jaw length, a trend which reached its greatest significance when canine area was scaled against mandibular length. When scaled to maxilloalveolar length, however, only males were positively allometric, females being nearly isometric. Previous studies on the linear correlation between canine size and body size indicated similar patterns of sexual dimorphism (Garn *et al.*, 1968a; Henderson & Corruccini, 1976; Anderson *et al.*, 1977).

The usefulness of the postcanine surface as a unit of comparison has repeatedly been demonstrated in primates (reviewed by Kieser & Groeneveld, 1987d). For example, when analysing ontogenetic scaling in rhesus monkeys, Cochard (1985) found that adult females had larger postcanine teeth for their body size than males. Similar studies in Sulawesi macaques (Albrecht, 1978) and in other Old World monkeys (Kay, 1978) suggested to Cochard (1985) that there may be a functional relationship between enlarged postcanine surface and body size during pregnancy. Observing a similar trend in Aboriginals, Wolpoff (1985) concluded that this may in fact be an evolutionary consequence of the elevated masticatory demands placed on the postcanine segments of gravid females. Whether viewed collectively or individually, the postcanine teeth of Negroes scaled negatively to jaw length. Though not wishing to interpret the orofacial complex as a mere accumulation of separate environmental adaptations, one needs to recognise the dependence of all its constituents on energy costs (see Peters, 1983). Rather than explain structural reduction in terms of 'non-adaptive' allometric trends (e.g. Gould & Lewontin, 1979), it is suggested here that the relatively smaller postcanine teeth in Negro males may be seen as an inevitable sequel to a reduction of energy expense.

There are, however, further possible interpretations of these results. Analysing relative incisor size in 57 species of anthropoids, Hylander (1975) argued that the relatively smaller incisors of man may be attributable to human food preparation techniques that have resulted in less incisal preparation of foodstuffs. But Kieser & Groeneveld (1988g) found incisor area to be unrelated to maxillary width, the positive correlations between incisor area and the three other viscerocranial measures being confined to males. Interestingly, Hylander (1975) based his analysis only on males. Clearly, further investigations into the allometry of incisors are called for.

Taken together, these craniodental allometric results indicate that the covariation between tooth size and viscocranial size is not particularly strong. This observation is further supported by craniodental allometric studies in the rhesus monkey (*Macaca mulatta*) (Cochard, 1985), *Cercopithecus aethiops* and *Papio ursinus* (Kieser & Groeneveld, 1987d, 1988a). Several explanations are possible, but the most plausible seems to be concerned with the early cessation of dental growth. Although teeth and jaws share a period of rapid size increase during foetal and early postnatal growth, subsequent growth will involve largely viscerocranial enlargement (tooth crowns having already reached their mature size). Hence the observed instances of tooth–jaw size covariation will be the result of growth determinants operative during foetal and early postnatal growth. Later growth will contribute to decreased tooth–jaw size covariation, since it affects viscerocranial rather than dental size. A similar distinction between the effects of early and late components of growth was drawn by Riska & Atchley (1985) in their analysis of brain–body size associations. There can be little doubt that the early cessation of dental growth affects also the general relationship betwen static and ontogenetic coefficients of allometry for craniodental traits. Although Cock (1966) emphasised the dangers of equating these two types of analyses, significantly positive associations between static and growth allometric coefficients are documented (Larson, 1981; Cheverud, 1982b; Leamy & Bradley, 1982; Cochard, 1985). Cock (1966), however, argued that the two kinds of coefficient can be related only if measured towards the end of a long period of growth, during which the ontogenetic development has been constant or nearly constant.

### 10.4.2    Dental allometry

The study by Kieser & Groeneveld (1988g) clearly indicated that when postcanine and incisor surfaces were scaled to canine base area, there were two definite allometric trends. The total postcanine surface scaled strongly negatively to canine area. That is, the larger the canine becomes, the smaller the posterior segments tended to be. The total incisor surface, on the other hand, scaled isometrically with canine area. Similarly, Kieser & Groeneveld (1987c) have shown that, when tooth size is corrected for arcade length, then the dentition may be divided into two negatively correlated segments: an anterior section consisting of canines and incisors, and a more distal postcanine section. Furthermore, available evidence suggests that the odontogenetic epithelium of the upper jaw is derived from four separate zones. Two zones are situated on the lateral borders of the frontonasal prominence and two on the maxillary process (Goose & Appleton, 1982). This suggests that the anterior teeth take their origin from the frontonasal rather than the maxillary prominence. From the foregoing,

Kieser & Groeneveld (1988g) concluded that a reduction in the size of canines and incisors will be accompanied by a disproportionate enlargement of premolar and molar size, a trend which might be echoed in the robust australopithecines.

### 10.4.3 *Allometry and evolutionary dwarfism*

Researchers have long speculated that dwarfed or pygmy forms should be characterised by strongly negative allometry. That is, dwarfed lineages should have disproportionately larger teeth than normally sized individuals (*vide* Gould, 1975). Despite frequent discussion of this problem, there has been little empirical study of the allometric relationships of tooth size in pygmy forms.

This issue was raised recently by Shea & Gomez (1988) who examined the allometric scaling of molar teeth in African and Philippine pygmies. As predicted by Gould (1975), these authors recorded a strongly negative allometric relationship between postcanine tooth size and skull size. The question then emerged, whether the relative enlargement of pygmy teeth, compared with those of larger humans, might be explained in terms of energetic or masticatory requirements. Earlier it was noted that, in some female primates, the postcanine teeth were found to be relatively larger than those of males of the same species (Cochard, 1985). This finding is usually explained in terms of the higher metabolic needs of pregnant and lactating females (e.g. Kieser & Groeneveld, 1987d, 1988a).

Shea & Gomez (1988), aware of the importance of a consideration of growth factors in craniodental scaling, proposed that the negative allometry might be due to differences in growth control substances. It was noted that the smaller stature of pygmies was the result of a genetically based inability to produce adequate levels of growth stimulating somatomedins. These insulin-like growth factors do not appear to affect the size of the dentition, hence pygmies might be expected to have larger teeth relative to their body sizes.

That the factors which control tooth size are predominantly independent of those that determine overall postnatal growth, was further borne out by the finding of very little covariation between tooth size and body size in pygmy humans (Shea & Gomez, 1988). This finding thus provides further support for the suggestion by Kieser & Groeneveld (1988g) that tooth size and body size are largely independently determined.

## 10.5 Conclusion

These excursions into allometric relations of the dentition reveal something of the complexity of the different interacting components involved in craniofacial development. This chapter has attempted to provide an introduction to several approaches to the ever widening field of allometry. Although many of the studies referred to in this section are less than conclusive, it is hoped that the reader will have gained some familiarity with an almost unexplored and highly important field of research.

An understanding of allometry is becoming increasingly important in the fields of human evolution and of clinical orthodontics. Past trends in allometric research, that will undoubtedly continue, have been toward a greater sophistication in experimental design and statistical evaluation. However, the aspiring researcher would do well to consider the basic problems faced by the authors mentioned in the text, prior to considering the more esoteric facets of allometry. For although allometry is a powerful tool for determining the effects of relative growth and size, its usefulness is determined by the clarity of the researcher's objectives and by the relevance of the statistical methods employed.

Before terminating the discussion on human craniodental allometry, one other interesting point should be made. It has become increasingly apparent recently that the patterns of covariation during ontogeny may place constraints upon morphological evolution (e.g. Smith *et al.*, 1985; Lande, 1986). Developmental interactions among certain characters may yield biased covariation patterns which, in turn, would restrain selection so as to act only on developmentally integrated characters. Moreover, the patterns of integration appear to vary throughout ontogeny (Atchley & Rutledge, 1980; Atchley, 1984). Hence, selection may be constrained by the fact that it can operate only at certain times during development. These times are, of course, those periods of covariation that characterise the ontogeny of the features under consideration.

An exciting attempt by Miriam Zelditch (1988) to deepen our insight into craniodental integration is a welcome one. Her investigation centred primarily on a study of five cross-sectional samples of a single population of laboratory rats at five age stages. It was found that the patterns of integration in the skull were stable throughout ontogeny, with only a temporary disruption between eye opening and weaning. From her analysis of craniodental integration, however, a different pattern emerged. First, there was a marked temporal and regional variation in the intensity of craniodental integration. Secondly, at certain age stages integration within the jaw deteriorated markedly. These results strongly suggest that at different periods during development, integration is determined by morphogenetic or functional or structural factors. This further suggests that the age at which selection acts may have a strong influence on the nature of the resultant morphological change. Zelditch (1988) noted, for example, that if selection were to act on weanling rats, the jaw and cranium could be affected separately. On the other hand, if selection acts upon the adult, it would modify the jaw with a concomitant effect on the neurocranium.

These results, together with those of the human craniodental allometric analyses mentioned earlier, present a serious challenge to the orthodox view of the functional and morphological integration of the orofacial region. It is hoped that this challenge will rekindle an interest in orofacial integration first demonstrated so strikingly in the seminal works of Moss and his associates.

# Appendix

*Mesiodistal and buccolingual diameters of maxillary and
mandibular dentition in human populations*

*Entry arrangement* population, reference, sex, sample size, mean, variability

*Sex*   M = male
     F = female
     B = both sexes pooled

*Variability*    '   variance
       *   standard error
    unmarked  standard deviation

M e s i o d i s t a l

**MAXILLA**

| | | I1 | I2 | C | P1 | P2 | M1 | M2 | M3 |
|---|---|---|---|---|---|---|---|---|---|
| Griqua (Campbell) (Kieser, 1985) | M | 12  8.26 0.68 | 12  6.90 0.62 | 15  8.07 0.37 | 17  6.99 0.81 | 14  6.64 0.52 | 18  10.77 0.61 | 15  10.12 0.65 | |
| | F | 9  7.92 0.35 | 11  6.47 0.50 | 10  7.65 0.13 | 13  6.56 0.47 | 10  6.37 0.47 | 13  9.89 0.18 | 14  9.15 0.56 | |
| Negro (W. Africa, Teso) (Barnes, 1969) | B | 27  8.85 0.56 | 28  7.16 0.42 | 17  7.97 0.32 | 14  7.54 0.67 | 10  7.40 0.41 | 13  10.63 0.61 | 13  9.78 0.53 | 16  9.13 0.77 |
| Negro (S. Africa) (Jacobson, 1982) | M | 144  8.82 0.66 | 195  7.14 0.68 | 270  7.91 0.43 | 295  7.27 0.43 | 294  6.80 0.38 | 292  10.64 0.53 | 303  10.32 0.71 | 265  9.24 0.73 |
| | F | 47  8.60 0.58 | 71  6.85 0.66 | 93  7.50 0.38 | 94  7.15 0.40 | 93  6.71 0.41 | 92  10.43 0.51 | 99  10.03 0.62 | 81  8.79 0.96 |
| Negro (S. Africa, contemporary) (Kieser et al., 1987) | M | 55  9.31 0.55 | 56  7.59 0.64 | 56  8.12 0.38 | 56  7.66 0.51 | 56  7.43 0.49 | 56  11.24 0.56 | 56  10.95 0.54 | |
| | F | 50  8.70 0.48 | 46  6.92 0.77 | 49  7.92 0.41 | 49  7.57 0.46 | 49  7.28 0.41 | 50  11.04 0.61 | 49  10.70 0.52 | |
| Negro (S. Africa, skeletal - post 1976) (Kieser et al., 1987) | M | 44  9.28 0.65 | 50  7.57 0.69 | 46  8.14 0.37 | 50  7.71 0.51 | 50  7.35 0.49 | 50  11.24 0.51 | 49  10.86 0.54 | |
| | F | 48  8.70 0.58 | 48  6.96 0.74 | 49  7.88 0.47 | 49  7.57 0.44 | 48  7.29 0.49 | 47  10.95 0.60 | 49  10.64 0.60 | |
| Nubian Mesolithic (10000-7000BC) (Calcagno, 1986) | M | 12  9.33 0.78 | 13  7.47 0.51 | 14  8.18 0.39 | 13  7.63 0.30 | 16  6.98 0.28 | 16  11.08 0.67 | 18  10.59 0.54 | 18  9.16 0.81 |
| | F | 14  9.21 0.38 | 12  7.15 0.59 | 14  7.79 0.52 | 15  7.37 0.46 | 15  6.75 0.42 | 16  10.75 0.51 | 18  10.34 0.62 | 18  8.96 0.59 |
| Nubian Agriculturalist (3300-1100BC) (Calcagno, 1986) | M | 15  8.90 0.71 | 25  6.85 0.51 | 43  7.68 0.44 | 55  7.17 0.43 | 77  6.80 0.40 | 71  10.49 0.53 | 91  9.79 0.65 | 88  8.72 0.73 |
| | F | 23  8.64 0.46 | 31  6.82 0.52 | 56  7.41 0.47 | 62  6.87 0.48 | 69  6.54 0.37 | 80  10.24 0.54 | 93  9.55 0.67 | 81  8.44 0.66 |
| Nubian Intensive Agriculturalist (1-1400AD) (Calcagno, 1986) | M | 14  8.91 0.62 | 23  7.10 0.79 | 40  7.83 0.37 | 47  7.09 0.40 | 51  6.75 0.35 | 55  10.42 0.39 | 58  9.73 0.57 | 53  11.05 0.99 |
| | F | 11  8.46 0.47 | 16  6.70 0.51 | 26  7.32 0.38 | 35  6.94 0.40 | 38  6.64 0.34 | 44  10.08 0.60 | 52  9.22 0.69 | 50  8.24 0.77 |
| San (Bushman, skeletal) (Drennan, 1929) | B | 25  8.3 | 24  6.7 | 26  7.5 | 27  6.8 | 28  6.5 | 33  9.9 | 30  9.7 | 26  8.2 |
| San (Bushman, skeletal) (van Reenen, 1982) | M | | 11  8.8 0.93 | 22  7.5 0.64 | 31  6.5 0.53 | 32  6.5 0.51 | 24  10.7 0.56 | 78  9.9 0.66 | 76  8.6 0.92 |
| | F | 8  8.0 1.11 | | 10  7.3 0.42 | 18  6.9 0.36 | 24  6.5 0.49 | 18  10.6 0.57 | 46  9.9 0.67 | 31  8.4 0.82 |
| San (Bushmen, Kalahari nomadic) (van Reenen, 1966) | M | | | | | | 117  10.78 0.69 | 114  10.22 0.80 | 60  8.68 0.92 |
| | F | | | | | | 128  10.45 0.62 | 121  9.98 0.74 | 54  8.11 0.74 |
| San (Bushmen, Kalahari settlers) (van Reenen, 1966) | M | | | | | | 16  10.95 0.66 | 14  10.51 0.72 | 9  9.53 0.73 |
| | F | | | | | | 32  10.80 0.78 | 32  10.22 0.86 | 18  9.53 1.31 |

## MANDIBLE

| | I1 | I2 | C | P1 | P2 | M1 | M2 | M3 |
|---|---|---|---|---|---|---|---|---|
| | 9 5.01 / 0.43 | 12 5.67 / 0.55 | 11 7.38 / 0.64 | 15 7.31 / 0.48 | 17 7.28 / 0.65 | 17 11.15 / 0.59 | 17 10.57 / 0.93 | |
| | 8 4.96 / 0.39 | 7 5.64 / 0.44 | 10 6.84 / 0.50 | 11 6.98 / 0.30 | 10 6.66 / 0.64 | 11 10.51 / 0.24 | 9 9.61 / 0.39 | |
| | 4 5.65 / 0.37 | 6 6.03 / 0.57 | 3 7.30 / 0.20 | 4 7.20 / 0.17 | 7 7.24 / 0.57 | 19 11.29 / 0.68 | 15 10.87 / 0.60 | 18 11.66 / 0.69 |
| | 181 5.33 / 0.42 | 223 6.06 / 0.47 | 264 7.33 / 0.45 | 300 7.38 / 0.46 | 293 7.44 / 0.54 | 277 11.39 / 0.60 | 284 11.08 / 0.74 | 278 11.02 / 0.87 |
| | 61 5.30 / 0.35 | 83 5.99 / 0.42 | 96 6.89 / 0.38 | 101 7.22 / 0.35 | 92 7.28 / 0.48 | 84 11.10 / 0.64 | 78 10.75 / 0.58 | 73 10.86 / 0.87 |
| | 56 5.43 / 0.45 | 52 6.43 / 0.51 | 56 7.57 / 0.47 | 56 7.88 / 0.46 | 55 7.92 / 0.51 | 54 11.81 / 0.47 | 55 11.55 / 0.60 | |
| | 49 5.48 / 0.56 | 48 6.11 / 0.55 | 48 7.31 / 0.38 | 49 7.64 / 0.50 | 44 7.80 / 0.54 | 48 11.60 / 0.59 | 46 11.28 / 0.54 | |
| | 46 5.56 / 0.44 | 45 6.41 / 0.48 | 48 7.59 / 0.49 | 45 7.86 / 0.53 | 47 7.90 / 0.48 | 47 11.84 / 0.48 | 43 11.47 / 0.61 | |
| | 49 5.46 / 0.53 | 48 6.12 / 0.57 | 49 7.24 / 0.45 | 48 7.61 / 0.50 | 47 7.72 / 0.51 | 47 11.52 / 0.67 | 44 11.14 / 0.61 | |
| | 8 5.81 / 0.35 | 14 6.31 / 0.53 | 15 7.51 / 0.41 | 13 7.50 / 0.40 | 17 7.37 / 0.38 | 14 11.85 / 0.49 | 16 11.39 / 0.58 | 17 11.18 / 0.79 |
| | 12 5.48 / 0.34 | 15 5.95 / 0.44 | 13 7.00 / 0.36 | 17 7.05 / 0.51 | 20 7.02 / 0.43 | 17 11.11 / 0.69 | 19 10.83 / 0.66 | 20 10.52 / 0.78 |
| | 26 5.44 / 0.35 | 35 6.04 / 0.31 | 38 6.87 / 0.50 | 75 7.13 / 0.45 | 86 7.25 / 0.41 | 81 11.41 / 0.60 | 104 10.93 / 0.68 | 105 10.60 / 0.85 |
| | 25 5.13 / 0.35 | 36 5.73 / 0.40 | 45 6.48 / 0.41 | 67 6.84 / 0.49 | 20 7.02 / 0.43 | 75 10.81 / 0.63 | 100 10.38 / 0.68 | 89 9.99 / 0.98 |
| | 16 5.44 / 0.26 | 25 6.09 / 0.34 | 34 6.94 / 0.35 | 50 7.02 / 0.44 | 55 7.09 / 0.51 | 53 11.26 / 0.48 | 61 10.76 / 0.68 | 54 10.37 / 0.86 |
| | 16 5.21 / 0.25 | 24 5.79 / 0.40 | 25 6.62 / 0.28 | 37 6.85 / 0.42 | 43 6.80 / 0.34 | 32 10.75 / 0.59 | 48 10.33 / 0.64 | 48 9.82 / 0.87 |
| | 22 5.0 | 24 5.6 | 29 6.8 | 27 6.9 | 30 7.0 | 33 10.9 | 32 10.6 | 33 9.9 |
| | 6 5.4 / 0.47 | 8 5.9 / 0.50 | 15 6.9 / 0.64 | 23 6.9 / 0.65 | 20 6.5 / 0.70 | 13 9.0 / 0.76 | 55 10.7 / 0.67 | 75 10.1 / 0.79 |
| | 5 5.4 / 0.84 | 7 5.7 / 0.25 | 13 6.9 / 0.55 | 25 6.7 / 0.46 | 23 6.9 / 0.83 | 6 10.9 / 0.83 | 34 10.5 / 0.74 | 38 9.4 / 0.98 |
| | | | | | | 118 10.85 / 0.69 | 118 10.62 / 0.67 | 91 10.04 / 0.83 |
| | | | | | | 130 10.74 / 0.71 | 126 10.74 / 0.66 | 74 9.80 / 0.91 |
| | | | | | | 14 9.89 / 1.29 | 12 10.63 / 0.51 | 16 9.94 / 1.54 |
| | | | | | | 27 10.59 / 0.81 | 29 10.43 / 0.94 | 21 9.96 / 1.10 |

# Buccolingual

## MAXILLA

| | | I1 | I2 | C | P1 | P2 | M1 | M2 | M3 |
|---|---|---|---|---|---|---|---|---|---|
| Griqua (Campbell) (Kieser, 1985) | M | 12 6.60<br>0.35 | 15 6.32<br>0.56 | 19 8.08<br>0.52 | 19 8.98<br>0.82 | 16 8.94<br>0.55 | 18 10.71<br>0.40 | 17 10.79<br>0.61 | |
| | F | 13 6.25<br>0.41 | 13 5.72<br>0.50 | 12 7.53<br>0.37 | 14 8.58<br>0.42 | 15 8.09<br>0.37 | 16 10.13<br>0.30 | 13 9.90<br>0.45 | |
| Negro (W. Africa, Teso) (Barnes, 1969) | B | 28 7.25<br>0.53 | 28 6.48<br>0.39 | 15 8.43<br>0.36 | 14 9.89<br>0.66 | 10 9.81<br>0.52 | 13 11.19<br>0.55 | 13 11.49<br>0.58 | 16 11.41<br>1.15 |
| Negro (S. Africa) (Jacobson, 1982) | M | 144 7.31<br>0.45 | 195 6.73<br>0.54 | 270 8.72<br>0.59 | 295 9.69<br>0.44 | 294 9.61<br>0.77 | 292 11.63<br>0.59 | 303 11.93<br>0.69 | 265 11.64<br>0.78 |
| | F | 47 7.14<br>0.50 | 71 6.42<br>0.51 | 93 8.10<br>0.57 | 94 9.34<br>0.52 | 93 9.36<br>0.58 | 92 11.32<br>0.60 | 99 11.52<br>0.67 | 81 10.84<br>1.13 |
| Negro (S. Africa, contemporary) (Kieser et al., 1987) | M | 55 7.56<br>0.54 | 55 7.10<br>0.57 | 56 8.87<br>0.61 | 56 9.83<br>0.52 | 56 9.82<br>0.67 | 56 11.80<br>0.64 | 56 11.90<br>0.45 | |
| | F | 46 7.39<br>0.43 | 46 6.62<br>0.63 | 45 8.44<br>0.49 | 46 9.46<br>0.61 | 43 9.26<br>0.68 | 46 11.56<br>0.54 | 44 11.66<br>0.54 | |
| Negro (S. Africa, skeletal - post 1976) (Kieser et al., 1987) | M | 48 7.59<br>0.49 | 45 7.14<br>0.53 | 47 8.87<br>0.61 | 46 9.91<br>0.52 | 49 9.93<br>0.60 | 48 11.76<br>0.50 | 44 11.93<br>0.55 | |
| | F | 43 7.33<br>0.40 | 50 6.63<br>0.55 | 49 8.49<br>0.58 | 49 9.59<br>0.63 | 45 9.43<br>0.68 | 44 11.51<br>0.50 | 41 10.56<br>0.71 | |
| Nubian Mesolithic (10000-7000BC) (Calcagno, 1986) | M | 14 7.60<br>0.27 | 14 7.00<br>0.34 | 15 8.74<br>0.52 | 16 10.09<br>0.45 | 17 10.32<br>0.44 | 15 12.60<br>0.41 | 19 12.75<br>0.54 | 18 12.01<br>0.67 |
| | F | 14 7.44<br>0.58 | 16 6.83<br>0.58 | 16 8.50<br>0.57 | 14 9.91<br>0.43 | 15 10.02<br>0.59 | 14 12.10<br>0.60 | 18 11.97<br>0.82 | 18 11.47<br>0.77 |
| Nubian Agriculturalist (3300-1100BC) (Calcagno, 1986) | M | 22 7.15<br>0.27 | 29 6.48<br>0.35 | 43 8.39<br>0.48 | 57 9.53<br>0.51 | 73 9.63<br>0.63 | 70 11.78<br>0.49 | 88 11.80<br>0.66 | 86 11.21<br>0.72 |
| | F | 25 6.89<br>0.34 | 36 6.31<br>0.43 | 56 7.86<br>0.57 | 58 9.16<br>0.59 | 69 9.24<br>0.59 | 77 11.46<br>0.55 | 89 11.36<br>0.66 | 83 10.82<br>0.79 |
| Nubian Intensive Agriculturalist (1-1400AD) (Calcagno, 1986) | M | 17 7.18<br>0.42 | 26 6.57<br>0.50 | 43 8.43<br>0.65 | 49 9.49<br>0.54 | 53 9.60<br>0.56 | 53 11.61<br>0.56 | 58 11.62<br>0.70 | 53 8.71<br>0.85 |
| | F | 16 7.19<br>0.35 | 24 6.25<br>0.35 | 29 7.96<br>0.44 | 38 9.15<br>0.52 | 38 9.23<br>0.54 | 47 11.24<br>0.55 | 56 10.97<br>0.56 | 51 10.61<br>0.81 |
| San (Bushman, skeletal) (Drennan, 1929) | B | 27 6.5 | 26 6.0 | 26 7.8 | 25 8.6 | 26 8.5 | 33 10.6 | 30 10.6 | 26 10.3 |
| San (Bushman, skeletal) (van Reenen, 1982) | M | | 17 6.2<br>0.57 | 22 7.6<br>0.57 | 31 8.4<br>0.55 | 32 8.5<br>0.55 | 24 11.1<br>0.55 | 78 11.0<br>0.64 | 32 10.5<br>0.66 |
| | F | 8 6.9<br>0.63 | 7 6.3<br>0.63 | 9 7.4<br>0.39 | 17 8.2<br>1.25 | 24 8.4<br>0.47 | 19 11.0<br>0.80 | 46 10.7<br>0.57 | 11 9.9<br>1.31 |
| San (Bushmen, Kalahari nomadic) (van Reenen, 1966) | M | | | | | | 117 10.96<br>0.72 | 118 10.96<br>0.76 | 78 10.53<br>0.92 |
| | F | | | | | | 129 10.91<br>0.62 | 124 10.88<br>0.60 | 65 10.21<br>0.80 |
| San (Bushmen, Kalahari settlers) (van Reenen, 1966) | M | | | | | | 16 11.21<br>0.35 | 14 11.32<br>0.56 | 31 10.57<br>1.37 |
| | F | | | | | | 31 10.92<br>0.74 | 30 10.90<br>0.72 | 19 10.34<br>0.79 |

## MANDIBLE

| I1 | I2 | C | P1 | P2 | M1 | M2 | M3 |
|---|---|---|---|---|---|---|---|
| 8 5.26 0.30 | 13 5.90 0.28 | 13 7.14 0.28 | 13 8.03 0.51 | 13 8.20 0.42 | 15 10.21 0.25 | 13 9.71 0.45 | |
| 6 5.15 0.11 | 8 5.72 0.24 | 6 6.62 0.36 | 7 7.17 0.25 | 16 7.33 0.33 | 10 10.06 0.27 | 6 9.52 0.48 | |
| 4 5.67 0.37 | 6 6.07 0.42 | 3 7.67 0.67 | 5 8.56 0.37 | 7 8.54 0.36 | 19 10.58 0.70 | 15 10.37 0.58 | 18 10.83 0.74 |
| 180 5.83 0.39 | 223 6.27 0.37 | 264 8.05 0.55 | 300 8.28 0.51 | 293 8.58 0.53 | 277 10.86 0.53 | 284 10.74 0.55 | 278 10.59 0.71 |
| 61 5.71 0.42 | 83 6.12 0.40 | 96 7.42 0.48 | 101 7.96 0.48 | 92 8.39 0.44 | 84 10.65 0.53 | 78 10.46 0.55 | 73 10.27 0.72 |
| 56 6.20 0.32 | 56 6.56 0.28 | 56 8.17 0.41 | 54 8.37 0.53 | 56 8.69 0.59 | 53 11.05 0.42 | 53 11.05 0.55 | |
| 45 5.96 0.49 | 46 6.45 0.46 | 46 7.48 0.45 | 46 7.82 0.40 | 46 8.30 0.61 | 45 11.77 0.56 | 40 10.71 0.62 | |
| 47 6.15 0.32 | 47 6.57 0.28 | 47 8.17 0.44 | 48 8.45 0.53 | 46 8.70 0.54 | 49 11.95 0.42 | 48 11.12 0.52 | |
| 47 5.89 0.43 | 49 6.37 0.44 | 49 7.48 0.47 | 47 8.23 0.58 | 46 8.70 0.58 | 44 10.95 0.42 | 40 10.82 0.58 | |
| 11 6.43 0.39 | 15 6.86 0.26 | 17 8.40 0.44 | 14 8.96 0.47 | 17 9.16 0.46 | 13 11.75 0.43 | 14 11.43 0.65 | 18 11.17 0.58 |
| 14 6.31 0.39 | 16 6.62 0.51 | 15 7.97 0.36 | 17 8.58 0.56 | 19 8.83 0.58 | 16 11.26 0.48 | 18 10.97 0.71 | 19 10.82 0.75 |
| 31 5.99 0.30 | 40 6.33 0.34 | 47 7.65 0.53 | 75 8.13 0.56 | 80 8.75 0.48 | 80 10.89 0.46 | 101 10.64 0.60 | 104 10.33 0.62 |
| 28 5.64 0.33 | 40 5.99 0.37 | 52 7.25 0.57 | 68 7.77 0.61 | 75 8.08 0.60 | 73 10.47 0.54 | 95 10.07 0.59 | 88 9.84 0.74 |
| 25 5.96 0.26 | 33 6.31 0.39 | 42 7.58 0.47 | 50 8.11 0.50 | 53 8.40 0.57 | 49 10.68 0.41 | 58 10.44 0.58 | 54 10.15 0.67 |
| 22 5.66 0.31 | 30 6.01 0.29 | 30 7.14 0.35 | 37 7.68 0.44 | 42 7.97 0.47 | 31 10.20 0.58 | 47 9.96 0.51 | 48 9.57 0.66 |
| 22 5.2 | 23 5.6 | 28 7.0 | 26 7.6 | 30 7.8 | 33 10.2 | 32 10.1 | 33 9.6 |
| 6 5.3 0.27 | 8 5.6 0.32 | 14 6.7 0.53 | 27 7.5 0.44 | 28 7.9 0.38 | 14 10.5 0.77 | 58 10.3 0.57 | 75 10.0 0.73 |
| 5 5.6 0.61 | 7 5.7 0.54 | 13 6.6 0.38 | 25 7.2 0.54 | 22 7.7 0.43 | 7 10.8 0.63 | 35 10.1 0.57 | 38 9.6 0.78 |
| | | | | | 118 10.52 0.48 | 118 10.43 0.57 | 95 10.17 0.74 |
| | | | | | 128 10.37 0.64 | 125 10.25 0.59 | 72 9.76 0.63 |
| | | | | | 14 10.27 1.06 | 14 10.77 0.63 | 16 10.01 1.11 |
| | | | | | 27 10.54 0.65 | 31 10.29 0.62 | 25 9.84 0.84 |

## Mesiodistal

### MAXILLA

| | | I1 | I2 | C | P1 | P2 | M1 | M2 | M3 |
|---|---|---|---|---|---|---|---|---|---|
| American (Hanihara, 1976) | M | 39  9.06 0.54 | 40  7.13 0.65 | 40  8.26 0.44 | 40  7.63 0.46 | 40  7.22 0.52 | 40 10.03 0.60 | 33 10.65 0.64 | |
| | F | 40  8.84 0.41 | 40  7.07 0.50 | 40  7.82 0.40 | 40  7.37 0.47 | 40  7.07 0.40 | 40 10.69 0.62 | 28 10.12 0.48 | |
| American (Hanihara & Ueda, 1979) | M | 37  9.07 0.56 | 37  9.13 0.67 | 37  8.25 0.44 | 37  7.63 0.46 | 37  7.21 0.54 | 37 11.03 0.62 | | |
| American (Richardson & Malhotra, 1975) | M | 153  9.12 0.67 | 159  7.26 0.64 | 142  8.19 0.53 | 146  7.66 0.49 | 148  7.25 0.49 | 148 10.04 0.64 | 104 10.74 0.63 | |
| | F | 156  8.72 0.58 | 156  7.08 0.56 | 145  7.74 0.38 | 148  7.37 0.43 | 138  6.94 0.39 | 140 10.57 0.52 | 112 10.35 0.73 | |
| British (Birmingham) (Lavelle, 1972) | M | 20  9.33 0.59 | 20  7.61 0.63 | 20  8.32 0.62 | 20  7.71 0.49 | 20  7.21 0.38 | 20 11.32 0.56 | | |
| | F | 20  9.21 0.44 | 20  7.48 0.38 | 20  8.21 0.52 | 20  7.63 0.96 | 20  7.18 0.15 | 20 10.09 0.48 | | |
| South American (Surinam) (Taverne, 1980) | M | 19  9.03 0.59 | 19  7.34 0.36 | 19  8.26 0.41 | 19  7.53 0.43 | 19  7.05 0.47 | 19 11.38 0.70 | 19 10.19 0.72 | |
| | F | 32  8.80 0.77 | 32  7.30 0.76 | 32  7.83 0.51 | 32  7.37 0.41 | 32  6.87 0.44 | 32 10.54 0.64 | 32  9.90 0.66 | |

## Buccolingual

### MAXILLA

| | | I1 | I2 | C | P1 | P2 | M1 | M2 | M3 |
|---|---|---|---|---|---|---|---|---|---|
| South America (Surinam) (Taverne, 1980) | M | | | | 19  9.97 0.44 | 19  9.82 0.56 | 19 11.79 0.65 | 19 11.87 0.57 | |
| | F | | | | 32  9.42 0.69 | 32  9.36 0.67 | 32 11.22 0.58 | 32 11.33 0.67 | |

MANDIBLE

| I1 | I2 | C | P1 | P2 | M1 | M2 | M3 |
|---|---|---|---|---|---|---|---|
| 40 5.60 | 40 6.20 | 39 7.31 | 40 7.77 | 40 7.75 | 39 12.01 | 27 11.47 | |
| 0.33 | 0.39 | 0.52 | 0.47 | 0.58 | 0.65 | 0.76 | |
| 40 5.46 | 40 6.01 | 40 6.82 | 40 7.44 | 40 7.60 | 40 11.47 | 22 10.89 | |
| 0.34 | 0.36 | 0.42 | 0.49 | 0.61 | 0.58 | 0.59 | |
| 37 5.59 | 37 6.20 | 37 7.31 | 37 7.78 | 37 7.77 | 37 12.00 | | |
| 0.33 | 0.40 | 0.53 | 0.49 | 0.60 | 0.66 | | |
| 158 5.53 | 159 6.13 | 148 7.37 | 140 7.76 | 147 7.85 | 136 11.76 | 80 11.53 | |
| 0.39 | 0.44 | 0.57 | 0.51 | 0.55 | 0.72 | 0.86 | |
| 160 5.38 | 161 5.99 | 158 6.86 | 156 7.41 | 143 7.61 | 127 11.28 | 102 10.94 | |
| 0.39 | 0.46 | 0.42 | 0.50 | 0.50 | 0.60 | 0.73 | |
| 20 5.83 | 20 6.59 | 20 7.44 | 20 7.45 | 20 7.54 | 20 11.87 | | |
| 0.55 | 0.41 | 0.57 | 0.38 | 0.64 | 0.44 | | |
| 20 5.81 | 20 6.49 | 20 7.41 | 20 7.44 | 20 7.51 | 20 11.61 | | |
| 0.72 | 0.39 | 0.66 | 0.63 | 0.54 | 0.75 | | |
| 18 5.54 | 18 6.27 | 18 7.39 | 18 7.41 | 18 7.46 | 18 11.58 | 18 10.97 | |
| 0.37 | 0.32 | 0.41 | 0.43 | 0.55 | 0.63 | 1.09 | |
| 29 5.51 | 29 6.11 | 29 6.87 | 29 7.30 | 29 7.35 | 29 11.24 | 29 10.60 | |
| 0.34 | 0.40 | 0.30 | 0.47 | 0.45 | 0.66 | 0.82 | |

MANDIBLE

| I1 | I2 | C | P1 | P2 | M1 | M2 | M3 |
|---|---|---|---|---|---|---|---|
| | | | 18 8.28 | 18 8.51 | 18 10.73 | 18 10.84 | |
| | | | 0.63 | 0.71 | 0.63 | 0.85 | |
| | | | 29 8.00 | 29 8.56 | 29 10.61 | 29 10.50 | |
| | | | 0.50 | 0.50 | 0.66 | 0.71 | |

M e s i o d i s t a l

| | | I1 | I2 | C | P1 | P2 | M1 | M2 | M3 |
|---|---|---|---|---|---|---|---|---|---|
| | | | | | | MAXILLA | | | |
| Pecos Pueblo (800-1100AD) (Nelson, 1938) | B | 105 8.67 0.35 | 84 7.06 0.38 | 86 8.05 0.45 | 82 7.43 0.39 | 88 7.01 0.47 | 67 10.74 0.48 | 79 9.94 0.47 | 84 9.36 0.60 |
| Indian Knoll (4160-2558BC) (Perzigian, 1976) | M | 30 8.97 0.47 | 32 7.55 0.52 | 43 8.27 0.44 | 33 7.20 0.52 | 24 6.89 0.37 | 35 11.11 0.66 | 38 10.02 0.49 | 29 9.01 0.76 |
| | F | 24 8.61 0.48 | 34 7.61 0.46 | 46 7.95 0.50 | 41 7.05 0.38 | 38 6.83 0.41 | 37 10.64 0.49 | 47 9.70 0.59 | 33 8.98 0.68 |
| Mexico, Cuaualan (mixed) (O'Rourke & Crawford, 1980) | M | 37 8.69 0.56 | 37 7.11 0.43 | 37 8.07 0.42 | 37 7.05 0.51 | 37 6.73 0.42 | 37 10.56 0.51 | | |
| | F | 40 8.50 0.51 | 40 7.14 0.56 | 40 7.68 0.32 | 40 6.78 0.46 | 40 6.68 0.41 | 40 10.29 0.47 | | |
| Mexico, Saltillo (mixed) (O'Rourke & Crawford, 1980) | M | 22 8.44 0.49 | 22 6.92 0.52 | 22 7.72 0.52 | 22 6.41 0.52 | 22 6.13 0.42 | 22 10.16 0.49 | | |
| | F | 64 8.12 0.65 | 64 6.50 0.54 | 64 7.37 0.43 | 64 6.23 0.49 | 64 6.19 0.55 | 64 9.76 0.62 | | |
| Mexico, San Pablo (O'Rourke & Crawford, 1980) | M | 31 8.55 0.44 | 31 6.88 0.67 | 31 8.02 0.53 | 31 6.69 0.49 | 31 6.47 0.53 | 31 10.34 0.45 | | |
| | F | 29 8.35 0.52 | 29 6.60 0.65 | 29 7.60 0.36 | 29 6.62 0.48 | 29 6.57 0.52 | 29 10.29 0.44 | | |
| Mexico, Tlaxcala (O'Rourke & Crawford, 1980) | M | 13 8.67 0.56 | 13 6.74 0.57 | 13 8.02 0.39 | 13 6.75 0.44 | 13 6.39 0.36 | 13 10.41 0.71 | | |
| | F | 29 8.35 0.52 | 29 6.60 0.65 | 29 7.60 0.36 | 29 6.62 0.48 | 29 6.57 0.52 | 29 10.29 0.44 | | |
| Ohio Valley (Adena) (Sciulli, 1979) | M | 18 8.89 0.47 | 22 7.35 0.57 | 23 8.53 0.51 | 19 7.09 0.45 | 19 6.71 0.35 | 20 10.57 0.71 | 21 9.93 0.55 | 17 8.43 0.89 |
| | F | 24 8.61 0.55 | 26 7.40 0.64 | 30 7.98 0.46 | 31 7.01 0.50 | 32 6.82 0.61 | 24 10.60 0.70 | 30 9.72 0.61 | 16 8.40 0.94 |
| Ohio Valley (Glacial Kame) (Sciulli, 1979) | M | 14 9.21 0.41 | 15 7.39 0.52 | 17 8.62 0.49 | 18 7.08 0.50 | 18 6.70 0.58 | 14 10.74 0.67 | 17 9.89 0.51 | 13 8.85 0.65 |
| | F | 14 8.80 0.40 | 13 7.50 0.54 | 17 8.17 0.35 | 16 7.19 0.43 | 18 6.75 0.36 | 19 10.53 0.51 | 19 9.66 0.59 | 16 8.77 0.62 |
| Ohio Valley (Hopewell) (Sciulli, 1979) | M | 15 8.77 0.52 | 18 7.46 0.33 | 22 8.40 0.44 | 21 7.32 0.31 | 19 6.82 0.49 | 18 10.90 0.64 | 19 10.31 0.40 | 16 8.77 0.73 |
| | F | 17 8.63 0.50 | 16 7.24 0.43 | 19 8.01 0.49 | 17 7.08 0.34 | 18 6.76 0.38 | 19 10.90 0.63 | 19 9.84 0.42 | 15 9.10 0.80 |
| Pima (Hanihara, 1976) | M | 60 9.14 0.38 | 60 7.68 0.37 | 60 8.78 0.48 | 60 7.80 0.46 | 60 7.45 0.45 | 60 11.03 0.54 | 60 10.46 0.53 | |
| | F | 60 8.72 0.43 | 60 7.47 0.45 | 60 8.24 0.34 | 60 7.48 0.40 | 60 7.21 0.42 | 60 10.71 0.48 | 60 10.26 0.48 | |
| Tennessee Archaic (6000-500BC) (Hinton et al., 1980) | B | 33 8.65 | 33 7.25 | 36 8.02 | 33 6.87 | 30 6.37 | 30 10.10 | 31 9.61 | 35 8.67 |
| Tennessee Woodland (700-1150AD) (Hinton et al., 1980) | B | 33 8.84 | 39 7.12 | 43 7.98 | 40 7.07 | 44 6.78 | 50 10.55 | 41 9.95 | 41 9.21 |
| Tennessee Mississippi (1300-1550AD) (Hinton et al., 1980) | B | 43 8.40 | 45 7.06 | 54 7.96 | 47 7.16 | 54 6.78 | 60 10.35 | 48 9.78 | 38 8.93 |
| Eskimo, Aleut (Moorees, 1957) | M | 97 8.45 0.05* | 88 7.29 0.05* | 81 8.03 0.04* | 77 7.15 0.04* | 62 6.65 0.06* | 53 10.37 0.10* | 51 10.00 0.10* | 27 9.16 0.17* |
| | F | 65 8.07 0.05* | 59 7.08 0.05* | 57 7.67 0.05* | 37 6.96 0.05* | 37 6.61 0.07* | 36 10.05 0.07* | 41 9.84 0.09* | 15 8.99 0.13* |
| Eskimo (Hall Beach) (Mayhall, 1979) | M | 37 8.49 0.60 | 37 7.21 0.50 | 36 7.99 0.50 | 36 7.21 0.40 | 30 6.73 0.50 | 33 11.05 0.50 | 33 10.26 0.50 | 20 9.06 1.20 |
| Eskimo (Igloolik) (Mayhall, 1979) | M | 138 8.70 0.50 | 125 7.20 0.70 | 119 8.20 0.50 | 110 7.30 0.50 | 105 6.80 0.50 | 116 11.10 0.60 | 90 10.40 0.60 | 39 9.20 1.10 |
| | F | 111 8.40 0.50 | 102 7.00 0.60 | 101 7.90 0.50 | 100 7.20 0.50 | 98 6.70 0.60 | 105 10.70 0.50 | 82 10.00 0.70 | 31 8.40 1.50 |

MANDIBLE*

| I1 | I2 | C | P1 | P2 | M1 | M2 | M3 |
|---|---|---|---|---|---|---|---|
| 84 5.52 | 74 6.14 | 94 7.27 | 105 7.15 | 91 7.42 | 67 11.96 | 62 11.41 | 62 11.13 |
| 0.26 | 0.28 | 0.44 | 0.40 | 0.46 | 0.55 | 0.45 | 0.65 |
| 31 5.66 | 44 6.52 | 39 7.48 | 32 7.23 | 35 7.37 | 49 11.82 | 36 11.29 | 33 11.23 |
| 0.42 | 0.46 | 0.40 | 0.47 | 0.52 | 0.46 | 0.32 | 0.91 |
| 29 5.56 | 37 6.41 | 40 7.01 | 39 6.95 | 39 7.11 | 46 11.43 | 43 10.98 | 26 10.82 |
| 0.27 | 0.33 | 0.36 | 0.43 | 0.48 | 0.53 | 0.55 | 0.91 |
| 37 5.44 | 37 6.18 | 37 7.05 | 37 7.07 | 37 7.22 | 37 11.36 | | |
| 0.37 | 0.34 | 0.48 | 0.46 | 0.47 | 0.61 | | |
| 40 5.52 | 40 6.02 | 40 6.66 | 40 6.77 | 40 6.99 | 40 10.07 | | |
| 0.59 | 0.38 | 0.39 | 0.35 | 0.43 | 0.52 | | |
| 22 5.10 | 22 5.76 | 22 6.87 | 22 6.56 | 22 6.80 | 22 10.84 | | |
| 0.47 | 0.53 | 0.60 | 0.48 | 0.48 | 0.35 | | |
| 64 5.03 | 64 5.73 | 64 6.33 | 64 6.34 | 64 6.67 | 64 10.37 | | |
| 0.45 | 0.48 | 0.38 | 0.58 | 0.64 | 0.58 | | |
| 31 5.26 | 31 5.95 | 31 6.95 | 31 6.75 | 31 6.97 | 31 11.28 | | |
| 0.49 | 0.49 | 0.44 | 0.51 | 0.47 | 0.63 | | |
| 29 5.23 | 29 5.82 | 29 6.61 | 29 6.65 | 29 6.78 | 29 10.93 | | |
| 0.35 | 0.38 | 0.38 | 0.52 | 0.51 | 0.49 | | |
| 13 5.23 | 13 5.87 | 13 6.93 | 13 6.55 | 13 7.05 | 13 11.29 | | |
| 0.47 | 0.43 | 0.48 | 0.57 | 0.46 | 0.52 | | |
| 29 5.23 | 29 5.82 | 29 6.61 | 29 6.65 | 29 6.78 | 29 10.93 | | |
| 0.35 | 0.38 | 0.38 | 0.52 | 0.51 | 0.49 | | |
| 14 5.11 | 16 6.07 | 16 7.46 | 17 7.10 | 18 7.04 | 18 11.35 | 26 11.27 | 16 10.44 |
| 0.54 | 0.39 | 0.47 | 0.49 | 0.47 | 0.65 | 0.80 | 0.96 |
| 24 5.16 | 31 6.15 | 30 7.06 | 31 6.88 | 33 6.96 | 33 11.29 | 31 10.91 | 26 10.25 |
| 0.46 | 0.45 | 0.45 | 0.50 | 0.55 | 0.52 | 0.68 | 0.87 |
| 15 5.00 | 18 6.05 | 18 7.39 | 18 6.72 | 19 6.86 | 20 11.70 | 24 11.28 | 16 10.94 |
| 0.45 | 0.55 | 0.51 | 0.57 | 0.55 | 0.54 | 0.57 | 0.83 |
| 17 5.32 | 19 6.24 | 18 7.15 | 18 6.88 | 19 6.96 | 20 11.55 | 20 11.01 | 15 10.14 |
| 0.43 | 0.68 | 0.41 | 0.53 | 0.48 | 0.65 | 0.67 | 0.71 |
| 19 5.22 | 21 6.13 | 25 7.63 | 25 7.26 | 26 7.54 | 27 11.96 | 26 11.60 | 21 11.21 |
| 0.29 | 0.43 | 0.46 | 0.51 | 0.50 | 0.46 | 0.46 | 0.80 |
| 14 5.41 | 18 6.27 | 20 7.06 | 19 7.00 | 22 7.24 | 22 11.39 | 23 11.07 | 21 10.78 |
| 0.42 | 0.37 | 0.46 | 0.40 | 0.46 | 0.51 | 0.69 | 1.01 |
| 60 5.89 | 60 6.71 | 60 7.77 | 60 7.61 | 60 7.67 | 60 11.80 | 60 11.47 | |
| 0.43 | 0.39 | 0.40 | 0.42 | 0.54 | 0.49 | 0.55 | |
| 60 5.64 | 60 6.31 | 60 7.16 | 60 7.27 | 60 7.33 | 60 11.32 | 60 10.91 | |
| 0.30 | 0.36 | 0.30 | 0.40 | 0.50 | 0.51 | 0.55 | |
| 34 5.22 | 37 6.13 | 38 7.11 | 39 6.65 | 42 6.86 | 39 10.79 | 43 10.81 | 45 10.67 |
| 24 5.03 | 36 6.13 | 40 7.22 | 45 7.03 | 47 7.03 | 55 11.23 | 53 11.04 | 45 10.95 |
| 42 4.82 | 50 6.03 | 56 7.04 | 59 7.03 | 55 7.14 | 46 11.27 | 44 11.07 | 33 10.72 |
| 98 5.23 | 100 6.09 | 91 7.20 | 94 7.01 | 81 7.17 | 47 11.56 | 43 11.19 | 15 11.13 |
| 0.04* | 0.03* | 0.04* | 0.06* | 0.05* | 0.08* | 0.14* | 0.29* |
| 73 5.08 | 72 5.90 | 74 6.71 | 64 6.85 | 56 7.02 | 20 11.20 | 36 11.16 | 20 11.30 |
| 0.05* | 0.04* | 0.04* | 0.04* | 0.07* | 0.11* | 0.10* | 0.13* |
| 34 5.33 | 36 6.40 | 19 7.32 | 41 7.19 | 35 7.11 | 35 11.70 | 23 11.01 | |
| 0.40 | 0.50 | 0.50 | 0.50 | 0.90 | 0.60 | 1.20 | |
| 119 5.30 | 124 6.40 | 117 7.40 | 109 7.20 | 103 7.30 | 117 11.70 | 83 11.30 | 19 11.50 |
| 0.40 | 0.40 | 0.40 | 0.40 | 0.90 | 0.90 | 0.70 | 0.70 |
| 90 5.30 | 99 6.20 | 98 7.00 | 101 7.00 | 87 7.00 | 100 11.10 | 84 10.80 | 10 10.70 |
| 0.40 | 0.40 | 0.40 | 0.50 | 0.80 | 0.90 | 0.70 | 0.90 |

B u c c o l i n g u a l

### MAXILLA

| | | I1 | I2 | C | P1 | P2 | M1 | M2 | M3 |
|---|---|---|---|---|---|---|---|---|---|
| Pecos Pueblo (800-1100AD) (Nelson, 1938) | B | 105 7.30 0.35 | 84 6.46 0.38 | 86 8.36 0.53 | 82 9.89 0.55 | 88 9.18 0.43 | 67 11.54 0.41 | 79 11.24 0.63 | 84 10.5? 0.57 |
| Indian Knoll (4160-2558BC) (Perzigian, 1976) | M | 45 7.27 0.46 | 44 6.55 0.51 | 47 8.73 0.49 | 49 10.04 0.54 | 48 9.53 0.63 | 59 11.88 0.49 | 57 11.89 0.69 | 34 11.46 0.79 |
| | F | 45 7.18 0.38 | 43 6.61 0.44 | 54 8.29 0.47 | 52 9.67 0.48 | 52 9.33 0.56 | 56 11.47 0.50 | 57 11.40 0.49 | 36 11.08 0.74 |
| Mexico, Cuaualan (mixed) (O'Rourke & Crawford, 1980) | M | 37 7.19 0.71 | 37 6.35 0.51 | 37 8.13 0.58 | 37 9.54 0.68 | 37 9.49 0.62 | 37 11.42 0.62 | | |
| | F | 40 7.01 0.54 | 40 6.44 0.42 | 40 7.98 0.44 | 40 9.11 0.51 | 40 9.15 0.57 | 40 10.97 0.63 | | |
| Mexico, Saltillo (mixed) (O'Rourke & Crawford, 1980) | M | 22 7.21 0.71 | 22 6.21 0.53 | 22 8.20 0.70 | 22 9.11 0.77 | 22 9.06 0.66 | 22 11.21 0.65 | | |
| | F | 64 6.93 0.72 | 64 6.16 0.69 | 64 7.85 0.60 | 64 8.79 0.67 | 64 8.85 0.68 | 64 10.81 0.68 | | |
| Mexico, San Pablo (O'Rourke & Crawford, 1980) | M | 31 7.42 0.62 | 31 6.45 0.63 | 31 8.26 0.59 | 31 9.33 0.58 | 31 9.25 0.76 | 31 11.40 0.45 | | |
| | F | 29 7.05 0.52 | 29 6.16 0.62 | 29 7.98 0.56 | 29 8.95 0.70 | 29 9.10 0.67 | 29 11.14 0.58 | | |
| Mexico, Tlaxcala (O'Rourke & Crawford, 1980) | M | 13 7.22 0.49 | 13 6.47 0.78 | 13 8.56 0.48 | 13 9.15 0.67 | 13 9.12 0.82 | 13 11.24 0.56 | | |
| | F | 29 7.05 0.52 | 29 6.16 0.62 | 29 7.98 0.56 | 29 8.95 0.70 | 29 9.10 0.67 | 29 11.14 0.58 | | |
| Ohio Valley (Adena) (Sciulli, 1979) | M | 18 7.11 0.38 | 24 8.34 0.45 | 24 8.85 0.45 | 20 9.72 0.63 | 21 9.46 0.59 | 22 11.93 0.54 | 23 11.88 0.58 | 20 10.90 0.97 |
| | F | 26 6.82 0.41 | 29 6.17 0.40 | 32 8.05 0.41 | 34 9.41 0.58 | 34 9.07 0.51 | 26 11.52 0.49 | 33 11.45 0.53 | 18 10.20 1.27 |
| Ohio Valley (Glacial Kame) (Sciulli, 1979) | M | 17 7.26 0.30 | 16 6.64 0.41 | 21 8.88 0.39 | 20 9.78 - 0.57 | 19 9.49 0.57 | 16 12.21 0.47 | 20 12.18 0.42 | 16 11.28 0.87 |
| | F | 15 6.97 0.36 | 14 6.17 0.34 | 18 8.23 0.44 | 18 9.34 0.58 | 18 9.34 0.48 | 19 11.77 0.48 | 18 11.63 0.46 | 17 10.62 1.09 |
| Ohio Valley (Hopewell) (Sciulli, 1979) | M | 16 7.16 0.24 | 19 6.52 0.31 | 24 8.78 0.34 | 22 9.70 0.37 | 19 9.49 0.46 | 20 12.03 0.40 | 21 11.89 0.56 | 18 10.89 0.73 |
| | F | 17 6.80 0.38 | 15 6.05 0.40 | 21 8.00 0.41 | 18 9.16 0.43 | 21 9.04 0.35 | 21 11.44 0.52 | 21 11.27 0.52 | 18 10.56 0.82 |
| Tennessee Archaic (6000-500BC) (Hinton et al., 1980) | B | 35 7.3 0.4 | 35 6.6 0.5 | 37 8.4 0.5 | 34 9.6 1.0 | 33 9.5 0.5 | 31 11.9 0.6 | 40 11.8 0.7 | 36 11.1 0.8 |
| Tennessee Woodland (700-1150AD) (Hinton et al., 1980) | B | 41 7.1 0.5 | 42 6.3 0.6 | 47 8.5 0.6 | 44 9.7 0.7 | 45 9.6 0.7 | 40 11.91 0.6 | 43 11.6 0.7 | 42 11.0 0.8 |
| Tennessee Mississippi (1300-1550AD) (Hinton et al., 1980) | B | 45 7.0 0.5 | 47 6.3 0.5 | 55 8.2 0.7 | 48 9.5 0.5 | 54 9.4 0.5 | 60 11.5 0.6 | 48 11.2 0.5 | 36 10.3 1.1 |
| Eskimo, Aleut (Moorees, 1957) | M | | | 65 8.47 0.05* | 81 9.31 0.07* | 69 9.13 0.07* | 45 11.34 0.09* | 42 11.36 0.13* | 22 10.76 0.25* |
| | F | | | 44 8.15 0.06* | 29 9.18 0.04* | 30 9.03 0.13* | 20 11.21 0.10* | 27 11.28 0.12* | 9 10.70 0.09* |
| Eskimo (Hall Beach) (Mayhall, 1979) | M | 33 7.70 0.60 | 35 7.20 0.60 | 33 8.40 0.40 | 41 9.20 0.60 | 39 9.00 0.50 | 40 12.00 0.50 | 36 11.90 0.70 | 23 11.00 1.40 |
| Eskimo (Igloolik) (Mayhall, 1979) | M | 129 7.70 0.60 | 111 72.0 0.80 | 105 8.70 0.60 | 118 9.50 0.60 | 109 9.30 0.60 | 123 11.80 0.70 | 104 11.80 0.80 | 50 11.00 1.30 |
| | F | 104 7.50 0.60 | 95 6.90 0.60 | 96 8.30 0.50 | 107 9.20 0.70 | 100 9.10 0.70 | 110 11.40 0.50 | 83 11.30 0.70 | 34 10.40 1.60 |

## MANDIBLE

| I1 | I2 | C | P1 | P2 | M1 | M2 | M3 |
|---|---|---|---|---|---|---|---|
| 84 5.87 0.35 | 74 6.06 0.31 | 94 7.71 0.50 | 105 7.91 0.52 | 91 8.42 0.50 | 67 10.74 0.47 | 62 10.47 0.47 | 62 **10.10** 0.53 |
| 43 5.76 0.29 | 51 6.09 0.35 | 46 7.67 0.53 | 52 8.33 0.40 | 55 8.48 0.47 | 66 11.32 0.50 | 53 10.93 0.56 | 39 **10.71** **0.76** |
| 41 5.69 0.37 | 51 6.03 0.38 | 48 7.21 0.44 | 56 8.12 0.40 | 60 8.37 0.44 | 63 11.02 0.44 | 58 10.67 0.48 | 34 **10.82** 0.86 |
| 37 5.86 0.52 | 37 6.13 0.43 | 37 7.51 0.65 | 37 8.06 0.66 | 37 8.39 0.52 | 37 10.66 0.52 | | |
| 40 5.64 1.61 | 40 6.07 0.52 | 40 7.22 0.56 | 40 7.74 0.44 | 40 8.12 0.53 | 40 10.41 0.44 | | |
| 22 6.01 0.87 | 22 6.33 0.82 | 22 7.62 0.79 | 22 7.72 0.59 | 22 7.98 0.44 | 22 10.22 0.44 | | |
| 64 5.97 0.72 | 64 6.23 0.82 | 64 7.25 0.67 | 64 7.47 0.65 | 64 7.99 0.70 | 64 9.93 0.69 | | |
| 31 6.00 0.47 | 31 6.22 0.50 | 31 7.35 0.68 | 31 7.68 0.58 | 31 8.21 0.56 | 31 10.40 0.52 | | |
| 29 5.83 0.75 | 29 6.26 0.60 | 29 7.22 0.78 | 29 7.38 0.64 | 29 8.01 0.63 | 29 9.97 0.63 | | |
| 13 5.87 0.43 | 13 6.35 0.62 | 13 7.72 0.98 | 13 7.79 0.62 | 13 8.07 0.74 | 13 10.09 0.62 | | |
| 29 5.83 0.75 | 29 6.26 0.60 | 29 7.22 0.78 | 29 7.38 0.64 | 29 8.01 0.63 | 29 9.97 0.63 | | |
| 16 5.48 0.52 | 18 5.89 0.41 | 17 7.78 0.47 | 19 8.10 0.50 | 21 8.31 0.39 | 17 10.94 0.48 | 27 10.56 0.52 | 17 **10.38** 0.58 |
| 26 5.28 0.33 | 30 5.73 0.29 | 31 7.18 0.40 | 34 7.79 0.32 | 36 7.88 0.58 | 32 10.68 0.57 | 32 10.37 0.53 | 30 **10.14** 0.63 |
| 15 5.78 0.32 | 20 6.07 0.22 | 20 7.98 0.39 | 18 8.05 0.63 | 20 8.47 0.43 | 21 11.05 0.31 | 29 10.78 0.42 | 18 **10.29** 0.48 |
| 20 5.66 0.38 | 19 5.88 0.31 | 18 7.22 0.42 | 19 7.86 0.44 | 19 8.23 0.45 | 20 10.71 0.31 | 20 10.49 0.51 | 16 **10.14** 0.81 |
| 21 5.86 0.31 | 23 6.18 0.34 | 26 7.97 0.57 | 27 8.19 0.42 | 27 8.50 0.38 | 27 11.06 0.37 | 27 10.70 0.39 | 23 **10.68** 0.62 |
| 16 5.29 0.42 | 19 5.70 0.42 | 21 6.98 0.47 | 22 7.67 0.40 | 21 7.94 0.37 | 20 10.60 0.29 | 20 10.16 0.52 | 20 **10.35** 0.48 |
| 36 5.8 0.4 | 40 6.1 0.3 | 41 7.6 0.6 | 45 8.2 0.5 | 44 8.4 0.5 | 41 11.1 0.5 | 45 10.7 0.9 | 48 **10.7** 0.7 |
| 27 5.6 0.4 | 37 6.1 0.5 | 43 7.7 0.5 | 46 8.4 0.5 | 49 8.6 0.6 | 56 11.1 0.8 | 54 10.8 0.6 | 45 10.8 0.7 |
| 44 5.6 0.4 | 52 5.9 0.4 | 58 7.5 0.6 | 59 8.0 0.5 | 56 8.3 0.5 | 51 10.7 0.4 | 44 10.3 0.5 | 34 10.0 1.0 |
| | | 74 7.93 0.07* | 86 7.82 0.06* | 79 8.40 0.07* | 49 10.56 0.08* | 45 10.58 0.10* | 13 10.15 0.15* |
| | | 57 7.58 0.06* | 49 7.58 0.06* | 42 8.14 0.09* | 16 10.29 0.11* | 37 10.92 0.10* | 17 10.42 0.12* |
| 37 6.40 0.80 | 42 6.90 0.50 | 36 7.80 0.70 | 44 7.80 0.40 | 44 8.40 0.50 | 40 11.20 0.40 | 27 11.10 0.40 | 10 11.10 0.80 |
| 115 6.30 0.60 | 122 6.70 0.50 | 109 8.00 0.70 | 111 8.00 0.50 | 109 8.50 0.60 | 114 11.30 0.60 | 83 11.20 0.80 | 26 11.20 0.70 |
| 101 6.20 0.50 | 114 6.50 0.50 | 101 7.50 0.50 | 105 7.60 0.50 | 99 8.20 0.50 | 109 10.90 0.60 | 89 10.70 0.60 | 16 10.70 1.00 |

## Mesiodistal

### MAXILLA

| | | I1 | I2 | C | P1 | P2 | M1 | M2 | M3 |
|---|---|---|---|---|---|---|---|---|---|
| Lengua (Kieser et al., 1985d) | M | 67 9.10 0.41 | 68 8.02 0.58 | 74 9.05 0.32 | 80 7.88 0.44 | 80 7.35 0.47 | 77 11.28 0.55 | 75 10.85 0.58 | 26 10.07 0.63 |
| | F | 36 8.71 0.49 | 61 7.60 0.53 | 48 8.11 0.44 | 59 7.60 0.49 | 57 7.30 0.66 | 61 11.08 0.47 | 60 10.32 0.60 | 16 9.72 0.59 |
| Peruvian (4000-2500BP) (Scott, 1979) | B | 12 8.3 0.6 | 9 6.9 0.4 | 18 7.6 0.7 | 15 6.8 0.2 | 19 6.8 0.6 | 19 9.9 0.5 | 21 9.7 0.8 | 15 9.0 0.8 |
| Peruvian (6500-4000BP) (Scott, 1979) | B | 8 7.0 1.5 | 11 6.4 1.2 | 15 7.3 1.3 | 19 6.5 1.2 | 16 6.3 1.1 | 25 10.5 0.6 | 20 10.0 0.7 | 18 8.9 0.8 |
| Peruvian (2500-1000BP) (Scott, 1979) | B | 8 7.7 0.5 | 8 6.9 0.5 | 11 7.9 0.4 | 14 7.0 0.6 | 15 6.7 0.6 | 11 10.3 0.7 | 14 10.1 0.7 | 8 8.5 0.6 |
| Surinam (Taverne, 1980) | M | 28 8.62 0.61 | 28 7.12 0.69 | 28 8.53 0.50 | 28 7.35 0.55 | 28 7.03 0.47 | 28 10.97 0.63 | 28 9.92 0.86 | |
| | F | 31 8.28 0.52 | 31 6.94 0.65 | 31 8.10 0.49 | 31 7.23 0.52 | 31 6.96 0.45 | 31 10.83 0.73 | 31 9.86 0.61 | |
| Ticuna (Harris & Nweeia, 1980b) | M | 23 8.73 | 25 7.45 | 26 8.46 | 27 7.54 | 27 6.98 | 27 10.39 | 26 9.81 | |
| | F | 18 8.77 | 19 7.50 | 26 8.19 | 26 7.57 | 26 7.06 | 26 10.39 | 26 9.57 | |

## Buccolingual

### MAXILLA

| | | I1 | I2 | C | P1 | P2 | M1 | M2 | M3 |
|---|---|---|---|---|---|---|---|---|---|
| Lengua (Kieser et al., 1985d) | M | | | | 79 9.48 0.47 | 80 9.39 0.47 | 78 11.76 0.37 | 78 11.41 0.47 | 35 11.28 0.64 |
| | F | | | | 60 9.26 0.36 | 62 9.19 0.42 | 64 11.34 0.42 | 60 10.73 0.46 | 28 10.78 0.58 |
| Peruvian (4000-2500BP) (Scott, 1979) | B | 12 7.0 0.4 | 9 6.3 0.2 | 17 7.7 0.5 | 15 9.2 0.4 | 19 9.2 0.6 | 19 11.1 0.7 | 20 11.3 0.8 | 14 10.8 1.0 |
| Peruvian (6500-4000BP) (Scott, 1979) | B | 8 6.5 0.5 | 11 6.1 0.6 | 15 7.9 0.7 | 18 9.1 0.5 | 14 8.7 1.1 | 20 11.3 0.6 | 20 10.9 0.7 | 18 10.5 1.0 |
| Peruvian (2500-1000BP) (Scott, 1979) | B | 8 7.1 0.6 | 8 6.4 0.6 | 12 8.1 0.5 | 15 9.1 0.5 | 15 9.0 0.5 | 11 11.3 1.0 | 14 11.4 0.8 | 8 10.3 0.6 |
| Surinam (Taverne, 1980) | M | | | | 28 9.56 0.47 | 28 9.25 0.48 | 28 11.69 0.55 | 28 11.41 0.62 | |
| | F | | | | 31 9.41 0.55 | 31 9.28 0.71 | 31 11.49 0.68 | 31 11.20 0.69 | |
| Ticuna (Harris & Nweeia, 1980b) | M | 24 7.83 | 25 6.83 | 26 8.62 | 27 9.67 | 27 9.38 | 27 11.72 | 26 11.50 | |
| | F | 21 7.51 | 21 6.84 | 26 8.29 | 26 9.67 | 26 9.38 | 26 11.44 | 26 11.19 | |

## MANDIBLE

| I1 | I2 | C | P1 | P2 | M1 | M2 | M3 |
|---|---|---|---|---|---|---|---|
| 64 5.71 | 66 6.56 | 70 7.68 | 73 7.68 | 74 7.62 | 66 11.98 | 67 11.32 | 12 10.95 |
| 0.37 | 0.52 | 0.41 | 0.49 | 0.42 | 0.33 | 0.56 | 0.68 |
| 44 5.46 | 42 6.20 | 41 6.98 | 47 7.23 | 49 7.61 | 41 11.58 | 47 10.82 | 9 10.13 |
| 0.48 | 0.51 | 0.41 | 0.47 | 0.51 | 0.44 | 0.47 | 0.53 |
| 9 4.9 | 7 5.7 | 9 6.9 | 13 7.1 | 15 7.0 | 23 10.8 | 18 10.7 | 18 10.7 |
| 0.6 | 0.3 | 0.5 | 0.4 | 0.4 | 0.7 | 0.8 | 0.6 |
|  | 5 4.8 | 4 6.5 | 13 6.7 | 15 6.7 | 16 11.1 | 18 10.8 | 14 10.2 |
|  | 0.9 | 0.1 | 0.7 | 0.7 | 0.8 | 0.5 | 0.7 |
| 10 5.2 | 9 6.0 | 9 6.8 | 14 6.9 | 12 7.1 | 12 11.3 | 15 10.8 | 9 10.2 |
| 1.0 | 0.5 | 0.3 | 0.7 | 0.6 | 0.7 | 0.8 | 0.6 |
| 28 5.29 | 28 6.24 | 28 7.32 | 28 7.12 | 28 7.19 | 28 11.68 | 28 10.66 |  |
| 0.38 | 0.44 | 0.45 | 0.51 | 0.47 | 0.58 | 0.57 |  |
| 32 5.28 | 32 6.15 | 32 6.92 | 32 7.10 | 32 7.08 | 32 11.24 | 32 10.46 |  |
| 0.34 | 0.36 | 0.38 | 0.35 | 0.45 | 0.57 | 0.68 |  |
| 25 5.33 | 25 6.36 | 27 7.24 | 27 7.17 | 27 7.24 | 24 11.32 | 24 10.68 |  |
| 28 5.44 | 26 6.40 | 28 6.90 | 28 7.09 | 28 7.10 | 14 10.09 | 19 10.34 |  |

## MANDIBLE

| I1 | I2 | C | P1 | P2 | M1 | M2 | M3 |
|---|---|---|---|---|---|---|---|
|  |  |  | 78 8.10 | 78 8.34 | 79 11.04 | 77 10.57 | 29 10.58 |
|  |  |  | 0.41 | 0.41 | 0.42 | 0.42 | 0.38 |
|  |  |  | 54 7.90 | 54 8.36 | 51 10.84 | 49 10.22 | 25 10.32 |
|  |  |  | 0.45 | 0.57 | 0.41 | 0.37 | 0.50 |
| 9 5.4 | 7 5.6 | 9 7.0 | 13 7.7 | 15 8.0 | 23 10.7 | 19 10.4 | 19 10.5 |
| 0.4 | 0.5 | 0.5 | 0.3 | 0.4 | 0.5 | 0.7 | 0.7 |
|  | 5 5.3 | 4 6.6 | 11 7.3 | 14 7.8 | 15 10.6 | 17 10.3 | 15 9.8 |
|  | 0.5 | 0.2 | 0.5 | 0.6 | 0.7 | 0.6 | 0.9 |
| 8 6.0 | 7 6.0 | 11 6.8 | 14 7.7 | 12 8.0 | 12 10.8 | 15 10.6 | 9 10.2 |
| 0.6 | 0.4 | 0.3 | 0.6 | 0.4 | 0.8 | 0.8 | 0.3 |
|  |  |  | 29 7.89 | 29 8.26 | 29 10.81 | 29 10.64 |  |
|  |  |  | 0.47 | 0.45 | 0.63 | 0.52 |  |
|  |  |  | 32 7.68 | 32 8.30 | 32 10.62 | 32 10.52 |  |
|  |  |  | 0.48 | 0.44 | 0.60 | 0.70 |  |
| 24 5.87 | 25 6.23 | 27 7.45 | 27 7.86 | 27 8.19 | 24 10.51 | 24 10.32 |  |
| 23 5.86 | 25 6.28 | 27 7.24 | 28 7.80 | 28 8.07 | 14 11.22 | 20 10.06 |  |

M e s i o d i s t a l

|  |  | MAXILLA | | | | | | | |
|---|---|---|---|---|---|---|---|---|---|
|  |  | I1 | I2 | C | P1 | P2 | M1 | M2 | M3 |
| Afghanistan (Pashtun) (Sakai et al., 1971) | M | 64 8.49 0.57 | 67 6.77 0.52 | 71 7.82 0.44 | 73 6.83 0.51 | 71 6.53 0.47 | 71 10.31 0.53 | 68 9.14 0.71 |  |
| Afghanistan (Tajik) (Sakai et al, 1971) | M | 36 8.34 0.47 | 36 6.56 0.66 | 36 7.59 9.33 | 38 6.59 0.39 | 35 6.38 0.46 | 35 9.94 0.60 | 37 8.92 0.54 |  |
| Bhutan (Prakash et al., 1979) | M | 35 8.42 0.65 | 35 6.91 0.65 | 35 7.67 0.55 | 35 6.93 0.53 | 35 6.49 0.61 | 35 10.07 0.52 | 35 8.87 0.85 |  |
| India (Inagaon 1700-700BC) (Lukacs, 1985) | B | 25 8.88 0.58 | 24 7.35 0.55 | 20 7.83 0.63 | 16 7.04 0.51 | 15 6.90 0.83 | 35 10.62 0.62 | 20 9.77 0.56 | 8 9.08 0.54 |
| India (Ramapuram) (Walimbe, 1985) | M | 95 9.45 0.79 | 95 7.67 0.73 | 94 8.63 0.49 | 95 7.43 0.47 | 93 7.00 0.46 | 95 11.33 0.56 | 95 9.99 0.71 | 28 9.24 0.95 |
| India (Bronze Age Harappans) (Dutta, 1983) | M |  |  |  |  |  | 46 10.36 0.09* | 49 9.77 0.11* | 36 9.03 0.14* |
|  | F |  |  |  |  |  | 29 10.16 0.11* | 30 9.57 0.17* | 21 8.60 0.20* |
| Pakistan (British contemporary) (Radnzic, 1987) | M | 60 8.79 0.31' | 60 6.87 0.31' | 60 8.04 0.18' | 60 7.09 0.14' | 60 6.91 0.21' | 60 10.69 0.39' |  |  |
| Tibet (Sharma, 1983) | M | 33 8.21 0.09* | 33 6.46 0.09* | 33 7.28 0.11* | 33 6.88 0.08* | 33 6.73 0.10* | 33 10.67 0.08* | 25 9.41 0.10* | 27 8.12 0.14* |
|  | F | 36 7.67 0.09* | 36 6.02 0.08* | 36 7.18 0.07* | 36 6.67 0.10* | 36 5.90 0.09* | 32 9.88 0.10* | 24 8.41 0.11* | 28 6.70 0.15* |
| South-east (Java) (Mijsberg, 1931) | M | 129 8.6 0.06* | 131 7.0 0.06* | 136 8.0 0.05* | 133 7.5 0.05* | 129 7.0 0.05* | 129 10.8 0.06* | 125 10.0 0.06* | 96 9.2 0.10* |
|  | F | 35 8.2 0.12* | 38 6.7 0.10* | 42 7.7 0.08* | 39 7.3 0.07* | 40 6.9 0.09* | 41 10.5 0.11* | 39 9.6 0.14* | 31 9.1 |
| South-east (Java) (Brace, 1979) | B | 19 8.58 0.54 | 24 7.02 0.54 | 31 7.92 0.51 | 34 7.41 0.50 | 34 6.99 0.41 | 30 10.54 0.62 | 31 9.74 0.66 | 24 8.40 1.15 |
| South-east (Java) (Taverne, 1980) | M | 42 8.24 0.90 | 42 6.66 0.70 | 42 8.06 0.56 | 42 7.46 0.45 | 42 6.87 0.41 | 42 10.75 0.64 | 42 9.98 0.76 |  |
|  | F | 37 8.46 0.68 | 37 6.71 0.60 | 37 7.82 0.56 | 37 7.44 0.49 | 37 7.10 0.52 | 37 10.60 0.68 | 37 9.51 0.71 |  |
| South-east (Bronze Age Java) (Brace, 1979) | B | 10 8.66 0.39 | 11 6.82 0.64 | 10 7.87 0.56 | 11 7.56 0.43 | 12 7.31 0.42 | 13 10.72 0.51 | 13 10.05 0.64 | 9 9.19 0.88 |
| South-east (Philippines) (Potter et al., 1981) | M | 87 8.33 0.49 | 94 6.76 0.63 | 84 7.75 0.51 | 82 6.86 0.44 | 85 6.56 0.50 | 82 10.02 0.67 | 77 9.24 0.67 |  |
|  | F | 142 8.03 0.48 | 137 6.44 0.66 | 144 7.45 0.45 | 128 6.82 0.47 | 132 6.41 0.41 | 123 9.77 0.50 | 122 9.65 0.56 |  |
| South-east (Thai) (Brace, 1979) | B | 46 8.50 0.50 | 50 7.12 0.56 | 52 7.84 0.48 | 52 7.35 0.40 | 50 6.75 0.35 | 54 10.38 0.52 | 53 9.68 0.60 | 43 9.11 0.73 |
| South-east (Bronze Age Thai) (Brace, 1979) | B | 34 8.46 0.40 | 31 6.87 0.73 | 38 7.72 0.53 | 44 7.24 0.50 | 48 6.73 0.49 | 52 10.45 0.56 | 44 9.45 0.72 | 32 9.19 1.22 |
| Chinese (Bronze Age Shang Dynasty) (Brace, 1979) | B | 222 8.48 0.53 | 258 6.99 0.66 | 302 7.72 0.45 | 314 7.09 0.41 | 319 6.51 0.48 | 323 10.11 0.53 | 308 9.56 0.59 | 219 8.83 0.84 |
| Chinese (Hosaka, 1936; quoted by Jacobson '82) | M | 267 8.68 0.03* | 219 6.98 0.03* | 210 8.06 0.03* | 209 7.21 0.03* | 143 6.86 0.03* | 88 10.02 0.06* | 70 9.36 0.06* | 50 8.60 0.10* |

## MANDIBLE

| I1 | I2 | C | P1 | P2 | M1 | M2 | M3 |
|---|---|---|---|---|---|---|---|
| 54 5.36 | 67 6.02 | 72 6.93 | 73 6.88 | 68 6.97 | 49 10.97 | 63 10.18 | |
| 0.39 | 0.41 | 0.39 | 0.45 | 0.47 | 0.73 | 0.57 | |
| 36 5.12 | 34 5.74 | 37 6.73 | 37 6.64 | 34 6.93 | 27 10.64 | 35 9.82 | |
| 0.36 | 0.37 | 0.39 | 0.36 | 0.45 | 0.57 | 0.57 | |
| 35 5.22 | 35 5.87 | 35 6.94 | 35 6.97 | 35 6.85 | 35 10.65 | 35 10.08 | |
| 0.40 | 0.47 | 0.49 | 0.50 | 0.54 | 0.72 | 0.76 | |
| 20 5.76 | 21 6.24 | 23 6.84 | 17 6.99 | 18 7.24 | 30 11.43 | 18 10.72 | 11 10.89 |
| 0.42 | 0.47 | 0.39 | 0.57 | 0.59 | 0.62 | 1.03 | 0.83 |
| 95 6.31 | 95 6.92 | 95 7.63 | 95 7.62 | 95 7.14 | 95 11.61 | 95 10.85 | 24 10.98 |
| 0.44 | 0.50 | 0.56 | 0.52 | 0.58 | 0.60 | 0.64 | 0.94 |
| | | | | | 23 10.57 | 24 10.15 | 24 9.96 |
| | | | | | 0.14* | 0.17* | 0.19* |
| | | | | | 13 10.50 | 15 9.67 | 13 9.27 |
| | | | | | 0.11* | 0.17* | 0.17* |
| 60 5.55 | 60 6.11 | 60 7.19 | 60 7.34 | 60 7.36 | 60 11.12 | | |
| 0.10' | 0.14' | 0.26' | 0.29' | 0.15' | 0.30' | | |
| 33 5.20 | 33 5.56 | 33 6.35 | 33 7.00 | 33 7.15 | 30 11.25 | 23 9.98 | 26 8.00 |
| 0.11* | 0.12* | 0.11* | 0.10* | 0.09* | 0.09* | 0.11* | 0.16* |
| 36 4.92 | 36 5.52 | 36 6.47 | 36 6.38 | 36 6.58 | 30 10.55 | 24 9.41 | 27 8.04 |
| 0.09* | 0.10* | 0.08* | 0.08* | 0.10* | 0.10* | 0.13* | 0.20* |
| 129 5.5 | 133 6.2 | 139 7.2 | 133 7.3 | 131 7.3 | 114 11.5 | 132 10.9 | 98 10.9 |
| 0.04* | 0.05* | 0.04* | 0.05* | 0.05* | 0.06* | 0.06* | 0.08* |
| 37 5.4 | 39 6.1 | 42 6.8 | 42 7.1 | 42 7.1 | 36 11.2 | 38 10.4 | 28 10.7 |
| 0.08* | 0.09* | 0.08* | 0.07* | 0.08* | 0.09* | 0.12* | 0.13* |
| 29 5.49 | 32 6.17 | 41 7.02 | 44 7.26 | 41 7.17 | 36 11.50 | 38 10.69 | 23 10.02 |
| 0.39 | 0.40 | 0.43 | 0.99 | 0.50 | 0.69 | 0.83 | 0.90 |
| 46 5.38 | 46 6.00 | 46 7.10 | 46 7.24 | 46 7.24 | 46 11.37 | 46 10.45 | |
| 0.52 | 0.43 | 0.47 | 0.40 | 0.43 | 0.59 | 0.64 | |
| 37 5.48 | 37 6.09 | 37 6.76 | 37 7.33 | 37 7.30 | 37 11.15 | 37 10.45 | |
| 0.40 | 0.38 | 0.41 | 0.53 | 0.52 | 0.52 | 0.81 | |
| 10 5.41 | 11 5.99 | 11 7.14 | 12 7.39 | 10 7.30 | 10 11.67 | 12 11.29 | 9 11.20 |
| 0.41 | 0.36 | 0.44 | 0.43 | 0.45 | 0.64 | 0.59 | 0.78 |
| 95 5.08 | 95 5.74 | 94 6.77 | 96 6.77 | 81 6.75 | 57 10.73 | 77 10.24 | |
| 0.34 | 0.39 | 0.47 | 0.44 | 0.47 | 0.68 | 0.78 | |
| 144 4.98 | 143 5.58 | 148 6.37 | 143 6.66 | 127 6.64 | 95 10.48 | 120 9.92 | |
| 0.34 | 0.43 | 0.41 | 0.50 | 0.54 | 0.64 | 0.72 | |
| 46 5.49 | 46 6.13 | 51 6.93 | 51 7.17 | 51 7.22 | 44 11.37 | 48 10.83 | 40 11.05 |
| 0.29 | 0.41 | 0.42 | 0.47 | 0.44 | 0.63 | 0.65 | 0.89 |
| 23 5.35 | 33 6.05 | 41 6.95 | 42 7.09 | 45 7.11 | 49 11.42 | 48 10.66 | 43 10.08 |
| 0.31 | 0.41 | 0.49 | 0.60 | 0.50 | 0.55 | 0.63 | 0.67 |
| 134 5.19 | 170 5.85 | 192 6.86 | 206 6.87 | 206 6.80 | 209 11.05 | 210 10.70 | 150 10.63 |
| 0.36 | 0.43 | 0.39 | 0.42 | 0.47 | 0.50 | 0.70 | 0.89 |
| 216 5.56 | 187 6.15 | 209 7.31 | 232 7.18 | 160 7.29 | 95 11.33 | 82 10.73 | 26 10.60 |
| 0.03* | 0.03* | 0.03* | 0.03* | 0.02* | 0.06* | 0.08* | 0.25* |

## M e s i o d i s t a l (*continued*)

**MAXILLA**

| | | I1 | I2 | C | P1 | P2 | M1 | M2 | M3 |
|---|---|---|---|---|---|---|---|---|---|
| Chinese (Hong Kong)<br>Brace, 1976) | B | 35  8.31<br>0.41 | 37  6.83<br>0.55 | 40  7.57<br>0.48 | 40  7.14<br>0.49 | 40  6.64<br>0.48 | 40  10.22<br>0.57 | 31  9.49<br>0.62 | 30  8.72<br>0.88 |
| Chinese (Hong Kong)<br>(Lavelle, 1972) | M | 20  8.67<br>0.42 | 20  6.99<br>0.77 | 20  8.07<br>0.66 | 20  7.19<br>0.55 | 20  6.81<br>0.52 | 20  9.99<br>0.59 | | |
| | F | 20  8.57<br>0.59 | 20  6.85<br>0.62 | 20  7.94<br>0.54 | 20  7.07<br>0.42 | 20  6.75<br>0.40 | 20  9.84<br>0.54 | | |
| Chinese (Liverpool)<br>(Lee & Goose, 1972) | M | 74  8.55<br>0.49 | 74  7.02<br>0.60 | 77  7.96<br>0.45 | 78  7.25<br>0.41 | 77  6.86<br>0.43 | 62  10.43<br>0.53 | 64  9.79<br>0.65 | |
| | F | 74  8.29<br>0.45 | 68  6.91<br>0.53 | 77  7.65<br>0.36 | 73  7.19<br>0.45 | 71  6.79<br>0.41 | 67  10.19<br>0.48 | 55  9.47<br>0.63 | |
| Chinese (Taiwan Atayal)<br>(Liu, 1977) | M | 151  8.7<br>0.03* | 151  7.1<br>0.05* | 152  8.1<br>0.04* | 152  7.4<br>0.03* | 147  7.0<br>0.03* | 152  10.8<br>0.05* | 114  9.9<br>0.06* | |
| | F | 80  8.6<br>0.06* | 78  6.7<br>0.06* | 83  7.8<br>0.05* | 81  7.2<br>0.05* | 86  6.9<br>0.04* | 78  10.9<br>0.08* | 50  9.6<br>0.09* | |
| Chinese (Taiwan Ami)<br>(Liu, 1977) | M | 131  8.5<br>0.05* | 132  7.2<br>0.05* | 135  8.1<br>0.04* | 137  7.4<br>0.03* | 133  6.9<br>0.03* | 138  10.5<br>0.04* | 127  9.9<br>0.05* | |
| | F | 83  8.3<br>0.04* | 74  6.8<br>0.06* | 83  7.8<br>0.04* | 83  7.2<br>0.05* | 82  6.6<br>0.05* | 80  9.9<br>0.06* | 73  9.6<br>0.07* | |
| Chinese (Taiwan Fujian)<br>(Yamada *et al.*, 1986) | M | 99  8.62<br>0.47 | 99  7.05<br>0.53 | 99  8.00<br>0.48 | 99  7.47<br>0.32 | 99  7.13<br>0.35 | 99  10.68<br>0.51 | 99  10.05<br>0.54 | |
| | F | 108  8.36<br>0.44 | 108  6.93<br>0.46 | 108  7.72<br>0.40 | 108  7.36<br>0.39 | 108  6.97<br>0.37 | 108  10.27<br>0.45 | 108  9.78<br>0.50 | |
| Japanese (Nagan)<br>(Suzuki & Sakai, 1957) | M | 100  8.68<br>0.36 | 97  7.18<br>0.39 | 108  8.22<br>0.17 | 108  7.40<br>0.22 | 105  4.94<br>0.22 | 103  10.68<br>0.37 | 108  9.69<br>0.43 | |
| | F | 35  8.55<br>0.40 | 35  7.15<br>0.40 | 35  7.91<br>0.42 | 35  7.62<br>0.41 | 35  6.97<br>0.41 | 35  10.33<br>0.45 | 35  9.55<br>0.56 | |
| Japanese (Hokoriku)<br>(Yamada 1932; quoted by Jacobson '82) | M | 132  8.5<br>0.03* | 127  7.1<br>0.03* | 142  7.8<br>0.03* | 140  7.3<br>0.03* | 140  6.8<br>0.03* | 151  10.2<br>0.03* | 142  9.8<br>0.03* | 76  8.8<br>0.08* |
| | F | 45  8.2<br>0.04* | 40  6.7<br>0.04* | 52  7.4<br>0.04* | 46  7.2<br>0.05* | 40  6.8<br>0.04* | 60  10.0<br>0.05* | 56  9.4<br>0.06* | 26  8.3<br>0.10* |
| Japanese (Ainu Hokkaido)<br>(Hanihara, 1976) | M | 20  8.38<br>0.42 | 20  7.18<br>0.37 | 20  7.91<br>0.32 | 20  6.99<br>0.38 | 20  6.47<br>0.33 | 20  10.40<br>0.45 | 20  9.08<br>0.56 | |
| | F | 14  8.28<br>0.44 | 14  6.88<br>0.78 | 14  7.62<br>0.45 | 14  6.98<br>0.35 | 14  6.31<br>0.51 | 14  10.02<br>0.44 | 14  9.11<br>0.70 | |
| Japanese (Ainu Hokkaido))<br>(Brace & Nagai, 1976) | M | 6  8.0<br>0.5 | 5  6.8<br>0.5 | 8  7.4<br>0.3 | 13  6.8<br>0.4 | 14  6.5<br>0.5 | 22  10.0<br>0.5 | 18  9.4<br>0.7 | 6  8.8<br>0.4 |
| | F | 4  7.8<br>0.2 | 6  6.9<br>0.3 | 9  7.6<br>0.4 | 10  6.8<br>0.3 | 15  6.2<br>0.3 | 20  10.0<br>0.4 | 12  9.2<br>0.4 | 9  8.6<br>0.7 |
| Japanese (Ainu Sakhalin)<br>(Suzuki & Sakai, 1957) | M | 10  8.62<br>0.29 | 10  7.65<br>0.10 | 10  8.36<br>0.24 | 10  7.36<br>0.19 | 10  6.94<br>0.14 | 8  10.38<br>0.25 | 7  9.52<br>0.21 | |
| | F | 8  8.35<br>0.02 | 8  7.50<br>0.12 | 8  7.93<br>0.09 | 8  7.34<br>0.10 | 8  6.98<br>0.03 | 8  10.53<br>0.14 | 7  9.35<br>0.25 | |

## MANDIBLE

| I1 | I2 | C | P1 | P2 | M1 | M2 | M3 |
|---|---|---|---|---|---|---|---|
| 29 5.11 | 39 5.79 | 41 6.79 | 41 6.82 | 36 6.84 | 29 10.64 | 27 10.43 | 12 10.16 |
| 0.43 | 0.40 | 0.43 | 0.55 | 0.66 | 0.58 | 0.85 | 0.81 |
| 20 5.52 | 20 6.09 | 20 7.41 | 20 7.20 | 20 7.25 | 20 11.28 | | |
| 0.89 | 0.47 | 0.54 | 0.52 | 0.59 | 0.62 | | |
| 20 5.49 | 20 5.86 | 20 7.34 | 20 7.11 | 20 7.18 | 20 11.09 | | |
| 0.41 | 0.43 | 0.47 | 0.54 | 0.56 | 0.63 | | |
| 73 5.47 | 76 6.03 | 74 7.04 | 74 7.11 | 75 7.18 | 50 11.23 | 58 10.41 | |
| 0.47 | 0.44 | 0.45 | 0.45 | 0.46 | 0.54 | 0.77 | |
| 73 5.37 | 78 5.96 | 78 6.64 | 73 7.05 | 62 7.06 | 49 10.94 | 60 10.12 | |
| 0.39 | 0.35 | 0.34 | 0.44 | 0.48 | 0.51 | 0.59 | |
| 153 5.5 | 151 6.3 | 150 7.0 | 156 7.0 | 152 7.2 | 132 11.3 | 129 10.7 | |
| 0.03* | 0.03* | 0.03* | 0.03* | 0.03* | 0.07* | 0.06* | |
| 79 5.4 | 82 5.9 | 83 6.7 | 84 6.9 | 83 7.1 | 69 11.2 | 70 10.2 | |
| 0.04* | 0.04* | 0.04* | 0.05* | 0.06* | 0.08* | 0.08* | |
| 132 5.6 | 127 6.2 | 137 7.1 | 183 7.1 | 135 7.1 | 113 11.3 | 131 10.3 | |
| 0.04* | 0.04* | 0.04* | 0.04* | 0.07* | 0.07* | 0.07* | |
| 81 5.3 | 81 6.0 | 80 6.7 | 84 6.9 | 84 6.8 | 72 10.9 | 83 10.0 | |
| 0.03* | 0.03* | 0.03* | 0.05* | 0.05* | 0.07* | 0.09* | |
| 99 5.52 | 99 6.12 | 99 7.02 | 99 7.27 | 99 7.38 | 99 11.41 | 99 10.66 | |
| 0.31 | 0.37 | 0.39 | 0.38 | 0.38 | 0.50 | 0.59 | |
| 108 5.44 | 108 5.99 | 108 6.67 | 108 7.21 | 108 7.27 | 108 11.01 | 108 10.28 | |
| 0.27 | 0.32 | 0.34 | 0.43 | 0.47 | 0.46 | 0.57 | |
| 100 5.65 | 97 6.28 | 108 7.21 | 108 7.28 | 105 7.13 | 103 11.53 | 108 10.80 | |
| 0.13 | 0.17 | 0.19 | 0.18 | 0.15 | 0.31 | 0.40 | |
| 35 5.65 | 35 6.19 | 35 6.89 | 35 7.34 | 35 7.19 | 34 11.16 | 35 10.60 | |
| 0.28 | 0.33 | 0.36 | 0.41 | 0.47 | 0.45 | 0.51 | |
| 129 5.4 | 136 6.0 | 146 7.0 | 143 7.1 | 138 7.1 | 137 11.2 | 141 11.0 | 88 10.2 |
| 0.03* | 0.03* | 0.02* | 0.05* | 0.03* | 0.03* | 0.04* | 0.06* |
| 44 5.2 | 51 5.9 | 48 6.6 | 49 7.0 | 47 7.1 | 52 10.9 | 45 10.6 | 26 9.5 |
| 0.04* | 0.04* | 0.04* | 0.03* | 0.04* | 0.05* | 0.08* | 0.10* |
| 20 5.38 | 20 6.01 | 20 7.09 | 20 6.95 | 20 6.69 | 20 11.23 | 20 10.45 | |
| 0.26 | 0.34 | 0.32 | 0.31 | 0.33 | 0.52 | 0.70 | |
| 14 5.30 | 14 5.81 | 14 6.59 | 14 6.92 | 14 6.74 | 14 10.80 | 14 10.14 | |
| 0.35 | 0.31 | 0.44 | 0.46 | 0.52 | 0.55 | 0.60 | |
| 5 5.2 | 9 5.6 | 11 6.5 | 12 6.5 | 14 6.8 | 14 11.1 | 12 10.5 | 9 10.2 |
| 0.2 | 0.2 | 0.4 | 0.3 | 0.5 | 0.6 | 0.5 | 0.4 |
| | | 6 6.5 | 6 6.6 | 9 6.5 | 14 10.7 | 11 10.0 | 8 9.7 |
| | | 0.3 | 0.4 | 0.2 | 0.4 | 0.5 | 0.6 |
| 8 5.69 | 9 6.16 | 9 7.45 | 9 7.12 | 8 7.04 | 6 11.48 | 7 10.52 | |
| 0.14 | 0.14 | 0.02 | 0.36 | 0.33 | 0.67 | 0.29 | |
| 8 5.59 | 8 6.13 | 8 7.03 | 8 7.21 | 8 7.13 | 8 11.08 | 7 10.05 | |
| 0.07 | 0.13 | 0.09 | 0.10 | 0.12 | 0.10 | 0.04 | |

# M e s i o d i s t a l (*continued*)

**MAXILLA**

| | | I1 | I2 | C | P1 | P2 | M1 | M2 | M3 |
|---|---|---|---|---|---|---|---|---|---|
| Japanese (Early Jomon) (Brace & Nagai, 1976) | M | 44  8.4<br>0.5 | 44  7.0<br>0.5 | 39  7.6<br>0.4 | 65  6.9<br>0.4 | 69  6.4<br>0.4 | 68  10.1<br>0.5 | 65  9.3<br>0.5 | 53  8.5<br>0.7 |
| | P | 16  8.2<br>0.4 | 17  6.7<br>0.4 | 14  7.0<br>0.4 | 34  6.6<br>0.4 | 43  6.2<br>0.4 | 43  9.8<br>0.5 | 34  9.0<br>0.5 | 53  8.3<br>1.0 |
| Japanese (Early Yoyoi) (Brace & Nagai, 1976) | M | 25  8.8<br>0.5 | 26  7.4<br>0.6 | 27  8.1<br>0.5 | 30  7.5<br>0.5 | 20  7.0<br>0.4 | 29  10.6<br>0.6 | 30  10.1<br>0.6 | 14  8.9<br>0.8 |
| | F | 14  8.5<br>0.4 | 13  7.2<br>0.4 | 8  7.9<br>0.4 | 13  7.2<br>0.5 | 13  6.8<br>0.5 | 16  10.3<br>0.5 | 10  9.8<br>0.6 | 6  8.9<br>0.4 |
| Japanese (Korean descent) (Brace & Nagai, 1976) | M | 12  8.7<br>0.3 | 11  7.3<br>0.4 | 16  7.9<br>0.4 | 24  7.3<br>0.4 | 23  6.9<br>0.4 | 30  10.2<br>0.5 | 30  9.8<br>0.5 | 12  8.8<br>0.4 |
| | F | | 6  6.6<br>0.5 | 9  7.7<br>0.3 | 12  7.2<br>0.3 | 9  6.9<br>0.3 | 14  10.1<br>0.4 | 14  9.6<br>0.5 | |
| Japanese (Kyoto) (Brace & Nagai, 1976) | M | 24  8.8<br>0.6 | 22  7.3<br>0.6 | 23  8.0<br>0.5 | 22  7.4<br>0.5 | 24  7.1<br>0.4 | 23  10.5<br>0.6 | 21  10.0<br>0.6 | 12  8.7<br>0.9 |
| | F | 17  8.2<br>0.5 | 18  6.7<br>0.6 | 16  7.4<br>0.4 | 18  7.1<br>0.4 | 16  7.0<br>0.4 | 18  10.1<br>0.6 | 18  9.4<br>0.6 | 17  8.6<br>1.1 |
| Japanese (Tokyo, Endo Period) (Brace & Nagai, 1976) | M | | 11  6.8<br>0.9 | 25  7.7<br>0.6 | 35  7.2<br>0.4 | 43  6.9<br>0.4 | 64  10.5<br>0.6 | 52  10.0<br>0.5 | 25  9.5<br>1.2 |
| | F | | | 5  7.9<br>0.4 | 10  7.4<br>0.5 | 13  6.8<br>0.4 | 21  10.4<br>0.7 | 21  9.9<br>0.7 | 4  9.7<br>0.7 |
| Japanese (Okinawa) (Mizoguchi, 1988) | M | 34  8.54<br>0.49 | 33  7.28<br>0.59 | 34  7.90<br>0.46 | 34  7.27<br>0.42 | 33  6.85<br>0.45 | 34  10.57<br>0.56 | 15  9.64<br>0.55 | |
| | F | 24  8.42<br>0.48 | 26  7.11<br>0.59 | 24  7.60<br>0.40 | 26  7.34<br>0.41 | 26  6.91<br>0.53 | 26  10.19<br>0.51 | 12  9.29<br>0.34 | |
| Japanese (Shimane) (Yamada et al., 1986) | M | 86  8.63<br>0.51 | 86  7.08<br>0.65 | 86  8.11<br>0.42 | 86  7.51<br>0.39 | 86  7.05<br>0.40 | 86  10.60<br>0.54 | 86  10.07<br>0.56 | |
| | F | 96  8.27<br>0.48 | 96  6.84<br>0.49 | 96  7.77<br>0.39 | 96  7.28<br>0.38 | 96  6.87<br>0,41 | 96  10.17<br>0.46 | 96  9.64<br>0.52 | |
| Japanese (Tokyo) (Mizoguchi, 1981b, 1986) | M | 51  8.64<br>0.53 | 51  7.12<br>0.50 | 51  8.16<br>0.35 | 51  7.47<br>0.39 | 51  6.96<br>0.42 | 51  10.59<br>0.50 | 51  9.77<br>0.58 | |
| | F | 142  8.45<br>0.49 | 142  6.94<br>0.61 | 137  7.71<br>0.45 | 142  7.36<br>0.45 | 120  6.86<br>0.39 | 137  10.31<br>0.49 | 89  9.62<br>0.47 | |
| Japanese (Tokyo, Hokkaido) (Hanihara, 1976) | M | 50  8.65<br>0.45 | 50  7.13<br>0.53 | 50  8.18<br>0.36 | 50  7.47<br>0.43 | 50  6.96<br>0.34 | 50  10.72<br>0.51 | 50  9.64<br>0.49 | |
| | F | 50  8.50<br>0.46 | 50  7.00<br>0.53 | 50  7.84<br>0.46 | 50  7.46<br>0.41 | 50  6.96<br>0.40 | 50  10.35<br>0.44 | 50  9.57<br>0.44 | |
| Japanese (Yamagata) (Sakai et al., 1971) | M | 91  8.50<br>0.52 | 91  7.10<br>0.54 | 91  8.06<br>0.46 | 89  7.52<br>0.42 | 90  7.16<br>0.40 | 86  10.78<br>0.58 | 83  9.89<br>0.65 | |

## MANDIBLE

| I1 | I2 | C | P1 | P2 | M1 | M2 | M3 |
|---|---|---|---|---|---|---|---|
| 26 5.2 | 35 5.8 | 62 6.7 | 74 6.8 | 80 6.8 | 84 11.4 | 85 10.8 | 71 10.6 |
| 0.3 | 0.3 | 0.2 | 0.4 | 0.4 | 0.6 | 0.6 | 0.8 |
| 14 5.1 | 19 5.6 | 25 6.4 | 39 6.6 | 53 6.7 | 57 11.1 | 61 10.4 | 44 10.2 |
| 0.3 | 0.3 | 0.2 | 0.4 | 0.5 | 0.6 | 0.7 | 0.9 |
| 23 5.5 | 32 6.2 | 40 7.2 | 40 7.3 | 38 7.5 | 32 11.8 | 32 11.4 | 26 11.0 |
| 0.3 | 0.4 | 0.4 | 0.5 | 0.5 | 0.5 | 0.7 | 0.8 |
| 9 5.4 | 11 6.0 | 13 6.9 | 14 7.0 | 14 7.0 | 14 10.3 | 13 10.7 | 8 10.6 |
| 0.2 | 0.2 | 0.4 | 0.5 | 0.5 | 0.4 | 0.6 | 0.8 |
| 11 6.5 | 10 6.2 | 11 7.0 | 10 7.3 | 12 7.1 | 13 11.4 | 13 11.3 | 9 11.0 |
| 0.3 | 0.2 | 0.4 | 0.4 | 0.4 | 0.5 | 0.4 | 0.6 |
| 4 5.4 | 5 5.9 | 5 6.8 | 4 7.2 | 4 7.3 | 13 11.0 | 6 10.7 | 6 9.9 |
| 0.2 | 0.3 | 0.2 | 0.1 | 0.2 | 0.4 | 0.8 | 1.1 |
| 20 5.5 | 21 6.2 | 20 7.2 | 22 7.3 | 21 7.3 | 20 11.5 | 20 11.3 | 14 10.5 |
| 0.4 | 0.4 | 0.3 | 0.4 | 0.5 | 0.5 | 0.7 | 0.7 |
| 17 5.3 | 17 5.9 | 17 6.6 | 17 7.1 | 16 7.3 | 16 11.1 | 13 10.9 | 10 10.2 |
| 0.3 | 0.4 | 0.4 | 0.4 | 0.4 | 0.5 | 0.7 | 1.6 |
| | 5 5.8 | 11 6.7 | 28 7.2 | 20 7.3 | 41 11.5 | 34 11.3 | 25 10.7 |
| | 0.9 | 0.5 | 0.4 | 0.4 | 0.7 | 0.8 | 0.8 |
| | | 4 6.4 | 6 6.8 | 9 6.9 | 19 11.1 | 16 10.9 | 7 10.5 |
| | | 0.4 | 0.5 | 0.6 | 1.0 | 0.9 | 0.8 |
| 32 5.38 | 34 5.98 | 34 6.95 | 34 7.09 | 32 7.15 | 28 11.53 | 32 10.50 | |
| 0.37 | 0.32 | 0.37 | 0.37 | 0.37 | 0.59 | 0.71 | |
| 25 5.36 | 23 5.93 | 24 6.64 | 25 7.12 | 24 7.21 | 23 11.20 | 14 10.58 | |
| 0.38 | 0.49 | 0.38 | 0.43 | 0.43 | 0.52 | 0.67 | |
| 86 5.47 | 86 6.13 | 86 7.06 | 86 7.47 | 86 7.51 | 86 11.75 | 86 11.32 | |
| 0.36 | 0.36 | 0.35 | 0.39 | 0.42 | 0.54 | 0.65 | |
| 96 5.34 | 96 5.93 | 96 6.75 | 96 7.16 | 96 7.16 | 96 11.17 | 96 10.56 | |
| 0.32 | 0.34 | 0.38 | 0.45 | 0.45 | 0.51 | 0.65 | |
| 51 5.44 | 51 6.15 | 51 7.19 | 51 7.40 | 51 7.26 | 51 11.54 | 51 10.85 | |
| 0.34 | 0.37 | 0.39 | 0.46 | 0.42 | 0.43 | 0.67 | |
| 122 5.37 | 127 5.99 | 141 6.71 | 141 7.19 | 121 7.17 | 135 11.26 | 61 10.43 | |
| 0.34 | 0.39 | 0.38 | 0.43 | 0.45 | 0.53 | 0.53 | |
| 50 5.56 | 50 6.17 | 50 7.10 | 50 7.30 | 50 7.19 | 50 11.49 | 50 10.56 | |
| 0.36 | 0.35 | 0.38 | 0.37 | 0.37 | 0.46 | 0.70 | |
| 48 5.45 | 49 6.10 | 50 6.85 | 50 7.25 | 50 7.18 | 50 11.27 | 50 10.42 | |
| 0.31 | 0.36 | 0.34 | 0.39 | 0.43 | 0.42 | 0.52 | |
| 84 5.52 | 88 6.17 | 90 7.18 | 91 7.49 | 83 7.50 | 76 11.63 | 76 11.16 | |
| 0.32 | 0.40 | 0.40 | 0.41 | 0.48 | 0.57 | 0.62 | |

B u c c o l i n g u a l

MAXILLA

| | | I1 | I2 | C | P1 | P2 | M1 | M2 | M3 |
|---|---|---|---|---|---|---|---|---|---|
| Afghanistan (Pashtun)<br>(Sakai et al., 1971) | M | 64  7.35<br>0.61 | 67  6.42<br>0.62 | 69  8.19<br>0.60 | 72  8.88<br>0.66 | 71  9.01<br>0.65 | 68  11.27<br>0.69 | 68  11.39<br>0.71 | |
| Afghanistan (Tajik)<br>(Sakai et al., 1971) | M | 36  7.31<br>0.62 | 36  6.52<br>0.70 | 36  8.21<br>0.71 | 38  8.86<br>0.77 | 35  8.79<br>0.73 | 35  11.14<br>0.58 | 38  11.26<br>0.66 | |
| Bhutan<br>(Prakash et al., 1979) | M | | | 35  7.52<br>0.89 | 25  9.18<br>0.78 | 35  9.05<br>0.79 | 35  11.16<br>0.81 | 35  11.11<br>0.88 | |
| India (Inagaon 1700-700BC)<br>(Lukacs, 1985) | B | 18  7.42<br>0.40 | 18  6.59<br>0.51 | 13  8.38<br>0.79 | 16  9.34<br>0.60 | 15  8.96<br>0.74 | 38  11.31<br>0.66 | 19  11.30<br>0.84 | 9  10.78<br>0.69 |
| India (Ramapuram)<br>(Walimbe, 1985) | M | 95  7.82<br>0.40 | 95  7.22<br>0.56 | 94  8.21<br>0.57 | 95  8.93<br>0.49 | 93  9.01<br>0.54 | 95  10.31<br>0.57 | 95  9.95<br>0.60 | 28  9.73<br>0.71 |
| India (Bronze Age Harappans)<br>(Dutta, 1983) | M | | | | | | 46  11.36<br>0.10* | 47  10.98<br>0.11* | 36  10.42<br>0.17* |
| | F | | | | | | 27  10.89<br>0.17* | 30  10.38<br>0.20* | 21  9.36<br>0.32* |
| Tibet<br>(Sharma, 1983) | M | | | | 33  7.00<br>0.08* | 33  7.22<br>0.06* | 33  9.15<br>0.11* | 25  8.37<br>0.12* | 27  7.82<br>0.15* |
| | F | | | | 36  6.74<br>0.11* | 36  6.86<br>0.11* | 32  8.65<br>0.07* | 24  8.07<br>0.14* | 28  6.97<br>0.21* |
| South-east (Java)<br>(Mijsberg, 1931) | M | 131  7.4<br>0.04* | 131  6.7<br>0.04* | 136  8.5<br>0.05* | 132  9.8<br>0.05* | 129  9.7<br>0.06* | 130  11.8<br>0.05* | 125  11.7<br>0.06* | 96  11.2<br>0.11* |
| | F | 35  6.9<br>0.09* | 38  6.3<br>0.09* | 43  7.8<br>0.07* | 39  9.4<br>0.09* | 40  9.2<br>0.09* | 41  11.2<br>0.08* | 39  10.9<br>0.12* | 31  10.7<br>0.12* |
| South-east (Java)<br>(Brace, 1979) | B | 19  7.26<br>0.47 | 24  6.59<br>0.44 | 31  8.30<br>0.56 | 34  9.78<br>0.57 | 34  9.60<br>0.53 | 30  11.39<br>0.68 | 21  11.34<br>0.85 | 21  10.70<br>1.35 |
| South-east (Java)<br>(Taverne, 1980) | M | | | | 42  9.66<br>0.64 | 42  9.52<br>0.64 | 42  11.38<br>0.70 | 42  11.49<br>0.76 | |
| | F | | | | 37  9.67<br>0.51 | 37  9.52<br>0.61 | 37  11.21<br>0.61 | 37  11.16<br>0.66 | |
| South-east (Bronze Age Java)<br>(Brace, 1979) | B | 9  7.42<br>0.44 | 11  6.64<br>0.71 | 10  8.24<br>0.68 | 11  9.76<br>0.52 | 12  9.51<br>0.52 | 13  11.73<br>0.52 | 13  11.64<br>0.60 | 9  11.33<br>0.52 |
| South-east (Philippines)<br>(Potter et al., 1981) | M | 87  7.00<br>0.71 | 94  6.19<br>0.71 | 84  7.94<br>0.67 | 82  9.42<br>0.53 | 85  9.28<br>0.59 | 82  11.11<br>0.64 | 77  11.04<br>0.71 | |
| | F | 142  6.84<br>0.62 | 136  6.08<br>0.66 | 144  7.55<br>0.60 | 128  9.25<br>0.51 | 132  9.10<br>0.55 | 123  10.79<br>0.54 | 122  10.74<br>0.68 | |
| South-east (Thai)<br>(Brace, 1979) | B | 46  7.25<br>0.42 | 50  6.53<br>0.41 | 52  8.24<br>0.57 | 52  9.52<br>0.62 | 50  9.38<br>0.57 | 54  11.44<br>0.50 | 53  11.34<br>0.66 | 43  11.04<br>0.39 |
| South-east (Bronze Age Thai)<br>(Brace, 1979) | B | 34  7.30<br>0.48 | 31  6.48<br>0.61 | 38  8.17<br>0.58 | 44  9.39<br>0.61 | 48  9.40<br>0.52 | 52  11.54<br>0.51 | 44  11.21<br>0.62 | 32  10.67<br>1.34 |
| Chinese (Bronze Age Shang Dynasty)<br>(Brace, 1979) | B | 222  7.19<br>0.43 | 258  6.46<br>0.54 | 302  8.35<br>0.53 | 314  9.46<br>0.51 | 319  9.26<br>0.55 | 323  10.38<br>0.52 | 308  11.34<br>0.62 | 219  10.80<br>0.96 |
| Chinese<br>(Hosaka, 1936; quoted by Jacobson '82) | M | | | 210  8.31<br>0.03* | 209  9.39<br>0.03* | 143  9.17<br>0.03* | 88  11.19<br>0.05* | 70  11.00<br>0.05* | 50  10.33<br>0.09* |
| Chinese (Hong Kong)<br>(Brace, 1976) | B | 35  7.21<br>0.39 | 37  6.48<br>0.47 | 40  8.25<br>0.61 | 40  9.40<br>0.52 | 40  9.22<br>0.60 | 31  11.24<br>0.55 | 30  11.22<br>0.81 | 31  10.30<br>1.18 |
| Chinese (Liverpool)<br>(Lee & Goose, 1972) | M | 71  7.45<br>0.51 | 71  6.83<br>0.57 | 76  8.51<br>0.60 | 76  9.51<br>0.58 | 74  9.41<br>0.52 | 61  11.40<br>0.57 | 63  11.44<br>0.65 | |
| | F | 73  7.19<br>0.42 | 69  6.52<br>0.48 | 79  8.01<br>0.51 | 74  9.21<br>0.52 | 75  9.10<br>0.54 | 69  10.94<br>0.47 | 57  10.95<br>0.66 | |
| Chinese (Taiwan Atayal)<br>(Liu, 1977) | M | | | 151  8.3<br>0.05* | 153  9.7<br>0.05* | 154  9.6<br>0.05* | 150  11.7<br>0.03* | 118  11.6<br>0.05* | |
| | F | | | 83  8.0 | 85  9.4 | 85  9.3 | 79  11.4 | 51  11.2 | |

MANDIBLE

| I1 | I2 | C | P1 | P2 | M1 | M2 | M3 |
|---|---|---|---|---|---|---|---|
| | | 49 7.85 0.51 | 73 7.69 0.58 | 68 9.10 0.55 | 51 10.36 0.68 | 61 10.11 0.66 | |
| | | 34 7.72 0.47 | 37 7.62 0.51 | 34 8.10 0.62 | 26 10.47 0.50 | 33 10.04 0.62 | |
| | | 35 6.97 1.04 | 35 7.94 0.48 | 35 8.25 0.52 | 35 10.64 0.74 | 35 10.33 0.71 | |
| 15 6.03 0.55 | 18 6.14 0.55 | 15 7.30 0.88 | 18 7.69 0.74 | 19 8.23 0.69 | 30 10.81 0.57 | 19 10.07 0.78 | 12 10.05 0.61 |
| 95 6.11 0.52 | 95 6.32 0.57 | 95 7.38 0.37 | 95 7.04 0.68 | 95 7.95 0.59 | 95 9.83 0.61 | 95 9.39 0.58 | 24 9.54 0.77 |
| | | | | | 23 10.54 0.18* | 24 9.88 0.25* | 24 9.44 0.22* |
| | | | | | 13 10.15 0.10* | 16 9.47 0.29* | 13 8.96 0.34* |
| | | | 33 5.88 0.11* | 33 6.21 0.09* | 30 9.20 0.11* | 23 8.01 0.12* | 25 7.05 0.15* |
| | | | 36 5.60 0.08* | 36 6.37 0.10* | 30 9.56 0.09* | 24 8.38 0.10* | 27 7.72 0.17* |
| 130 5.9 0.04* | 133 6.3 0.04* | 139 7.9 0.04* | 132 8.2 0.06* | 131 8.5 0.05* | 117 11.0 0.06* | 133 10.5 0.06* | 99 10.4 0.07* |
| 37 5.5 0.08* | 39 6.0 0.07* | 42 7.2 0.07* | 42 7.7 0.06* | 42 8.2 0.08* | 35 10.7 0.10* | 38 10.2 0.10* | 29 10.2 0.11* |
| 29 5.79 0.35 | 32 6.21 0.44 | 41 7.66 0.63 | 44 8.23 0.89 | 41 8.41 0.43 | 36 10.95 0.61 | 38 10.38 0.72 | 33 10.45 0.87 |
| | | | 42 8.28 0.69 | 42 8.72 0.58 | 42 10.90 0.65 | 42 10.63 0.62 | |
| | | | 37 8.03 0.52 | 37 8.68 0.47 | 37 10.67 0.51 | 37 10.56 0.65 | |
| 10 6.27 0.58 | 11 6.65 0.60 | 11 8.05 0.70 | 12 8.22 0.38 | 10 8.55 0.43 | 10 10.90 0.46 | 12 10.72 0.44 | 9 10.51 0.72 |
| 95 5.58 0.51 | 95 6.01 0.62 | 94 7.18 0.64 | 96 7.78 0.57 | 81 8.11 0.58 | 57 10.36 0.58 | 77 10.12 0.61 | |
| 144 5.47 0.42 | 142 5.84 0.43 | 148 6.84 0.53 | 143 7.53 0.50 | 126 7.97 0.56 | 95 10.15 0.61 | 120 9.86 0.63 | |
| 46 5.78 0.42 | 46 6.29 0.42 | 51 7.64 0.50 | 51 8.03 0.62 | 51 8.38 0.59 | 44 10.86 0.48 | 48 10.42 0.53 | 40 10.30 0.49 |
| 23 5.96 0.38 | 33 6.32 0.48 | 41 7.70 0.56 | 42 8.08 0.61 | 45 8.39 0.59 | 49 11.02 0.52 | 48 10.53 0.54 | 43 10.49 0.61 |
| 134 5.70 0.36 | 170 6.14 0.38 | 192 7.80 0.47 | 206 7.95 0.49 | 206 8.26 0.48 | 209 10.89 0.50 | 210 10.50 0.55 | 150 10.22 0.74 |
| | | 209 7.89 0.03* | 232 8.07 0.03* | 160 8.13 0.03* | 95 10.67 0.03* | 82 10.37 0.04* | 26 10.35 0.13* |
| 29 5.86 0.45 | 39 6.23 0.40 | 41 7.78 0.59 | 41 7.88 0.59 | 36 8.22 0.57 | 29 10.66 0.57 | 27 10.19 0.65 | 12 9.82 0.50 |
| 44 6.24 0.48 | 46 6.61 0.53 | 42 7.95 0.53 | 75 8.00 0.55 | 74 8.25 0.50 | 51 10.78 0.54 | 58 10.61 0.53 | |
| 59 6.03 0.46 | 62 6.45 0.43 | 67 7.47 0.56 | 74 7.72 0.48 | 64 8.11 0.21 | 50 10.40 0.46 | 59 10.24 0.53 | |
| | | 151 7.6 0.04* | 135 8.0 0.05* | 154 8.7 0.05* | 131 10.9 0.04* | 130 10.6 0.05* | |
| | | 81 7.4 | 84 8.0 | 80 8.3 | 69 10.6 | 75 10.4 | |

B u c c o l i n g u a l (*continued*)

MAXILLA

| | | I1 | I2 | C | P1 | P2 | M1 | M2 | M3 |
|---|---|---|---|---|---|---|---|---|---|
| Chinese (Taiwan Ami) (Liu, 1977) | M | | | | | 135 8.6 0.06* | 136 10.0 0.04* | 137 9.7 0.06* | 121 11.6 0.06 |
| | F | | | | | 84 8.3 0.07* | 84 9.8 0.06* | 84 9.2 0.06* | 84 11.3 0.07* |
| Chinese (Taiwan Fujian) (Yamada *et al.*, 1986) | M | 99 7.05 0.53 | 99 6.21 0.52 | 99 8.06 0.54 | 99 9.57 0.44 | 99 9.40 0.43 | 99 11.16 0.44 | 99 11.09 0.44 | |
| | F | 108 6.98 0.52 | 108 6.25 0.55 | 108 7.87 0.49 | 108 9.31 0.46 | 108 9.15 0.47 | 108 10.74 0.48 | 108 10.74 0.46 | |
| Japanese (Nagan) (Suzuki & Sakai, 1957) | M | | | | 107 9.61 0.40 | 102 9.55 0.40 | 100 11.74 0.42 | 108 11.76 0.33 | |
| | F | | | | 34 9.38 0.24 | 34 9.28 0.23 | 34 11.27 0.19 | 34 11.37 0.27 | |
| Japanese (Hokoriku) (Yamada, 1932; quoted by Jacobson '82) | M | | | 142 8.4 0.04* | 140 9.5 0.03* | 140 9.4 0.03* | 151 11.4 0.03* | 142 11.6 0.03* | 76 10.8 0.07* |
| | F | | | 52 7.9 0.04* | 46 9.3 0.06* | 46 9.2 0.04* | 60 11.0 0.04* | 56 10.9 0.04* | 26 10.0 0.11* |
| Japanese (Ainu Hokkaido) (Brace & Nagai, 1976) | M | 6 6.8 0.2 | 5 6.2 0.3 | 8 7.8 0.2 | 13 9.0 0.6 | 14 8.7 0.7 | 22 11.3 0.5 | 18 11.1 0.6 | 6 10.6 0.4 |
| | F | 4 6.6 0.4 | 6 6.2 0.3 | 9 7.3 0.5 | 10 8.8 0.5 | 15 8.4 0.5 | 20 10.9 0.4 | 12 10.5 0.8 | 9 10.1 0.7 |
| Japanese (Ainu Sakhalin) (Suzuki & Sakai, 1957) | M | | | | 10 9.39 0.29 | 10 9.10 0.33 | 8 11.55 0.10 | 7 11.32 0.20 | |
| | F | | | | 8 9.26 0.11 | 8 9.09 0.16 | 8 11.41 0.11 | 7 10.95 0.12 | |
| Japanese (Early Jomon) (Brace & Nagai, 1976) | M | 44 7.2 0.4 | 44 6.7 0.5 | 39 8.1 0.4 | 65 9.3 0.5 | 69 9.0 0.6 | 68 11.7 0.5 | 65 11.3 0.6 | 53 10.8 0.5 |
| | F | 16 6.9 0.4 | 17 6.4 0.4 | 14 7.4 0.5 | 34 9.0 0.6 | 43 3.8 0.5 | 43 11.3 0.5 | 34 10.9 0.6 | 53 10.3 1.2 |
| Japanese (Early Yoyoi) (Brace & Nagai, 1976) | M | 25 7.6 0.4 | 26 7.0 0.5 | 27 8.8 0.6 | 30 9.8 0.5 | 30 9.5 0.4 | 29 12.0 0.4 | 30 11.8 0.5 | 14 10.8 1.5 |
| | F | 14 7.1 0.5 | 13 6.6 0.4 | 8 8.1 0.4 | 13 9.3 0.6 | 13 9.2 0.6 | 13 11.5 0.6 | 16 11.3 0.5 | 10 10.4 0.5 |
| Japanese (Korean descent) (Brace & Nagai, 1976) | M | 12 7.6 0.4 | 11 6.9 0.5 | 16 8.7 0.8 | 24 9.6 0.7 | 23 9.6 0.6 | 30 11.6 0.5 | 30 11.8 0.7 | 12 11.0 0.8 |
| | F | | 6 6.4 0.3 | 9 8.0 0.6 | 12 9.2 0.4 | 9 9.2 0.6 | 14 11.5 0.5 | 14 11.3 0.6 | |
| Japanese (Kyoto) (Brace & Nagai, 1976) | M | 24 7.5 0.3 | 22 6.8 0.4 | 23 8.8 0.5 | 22 9.7 0.6 | 24 9.6 0.7 | 23 11.9 0.4 | 21 12.1 0.7 | 12 11.2 0.9 |
| | F | 17 7.0 0.4 | 18 6.5 0.4 | 16 8.0 0.5 | 18 9.2 0.5 | 16 9.2 0.5 | 18 11.2 0.5 | 18 11.0 0.9 | 7 10.4 1.7 |
| Japanese (Tokyo, Endo Period) (Brace & Nagai, 1976) | M | | 11 6.5 0.6 | 25 8.5 0.7 | 35 9.6 0.6 | 43 9.5 0.6 | 64 11.6 0.5 | 52 11.7 0.6 | 25 11.3 0.8 |
| | F | | | 5 8.6 0.6 | 10 9.5 0.5 | 13 9.2 0.5 | 21 11.4 0.7 | 21 11.5 1.0 | 4 11.6 0.7 |
| Japanese (Shimane) (Yamada *et al.*, 1986) | M | 86 7.43 0.50 | 86 6.56 0.58 | 86 8.60 0.60 | 86 9.73 0.53 | 86 9.48 0.54 | 86 11.86 0.51 | 86 12.03 0.66 | |
| | F | 96 7.03 0.53 | 96 6.32 0.45 | 96 8.08 0.45 | 96 9.42 0.51 | 96 9.21 0.54 | 96 11.22 0.46 | 96 11.32 0.59 | |
| Japanese (Yamagata) (Sakai *et al.*, 1971) | M | 32 7.30 0.54 | 37 6.58 0.63 | 71 8.27 0.79 | 89 9.70 0.47 | 88 9.63 0.65 | 85 11.65 0.57 | 72 11.95 0.56 | |

## MANDIBLE

| I1 | I2 | C | P1 | P2 | M1 | M2 | M3 |
|---|---|---|---|---|---|---|---|
| | | 136 7.9<br>0.05* | 135 8.4<br>0.05* | 131 8.6<br>0.04* | 111 10.8<br>0.05* | 128 10.6<br>0.05* | |
| | | 81 7.4<br>0.05* | 84 8.1<br>0.05* | 83 8.4<br>0.04* | 73 10.7<br>0.07* | 84 10.1<br>0.06* | |
| 99 5.93<br>0.37 | 99 6.30<br>0.36 | 99 7.72<br>0.51 | 99 8.05<br>0.48 | 99 8.44<br>0.43 | 99 10.76<br>0.41 | 99 10.52<br>0.48 | |
| 108 5.83<br>0.38 | 108 6.17<br>0.39 | 108 7.29<br>0.48 | 108 7.83<br>0.46 | 108 8.31<br>0.43 | 108 10.37<br>0.35 | 108 10.09<br>0.44 | |
| | | 107 8.13<br>0.32 | 102 8.51<br>0.26 | 100 10.98<br>0.31 | 108 10.57<br>0.31 | | |
| | | 34 7.77<br>0.11 | 34 8.27<br>0.19 | 33 10.67<br>0.11 | 34 10.26<br>0.21 | | |
| | | 146 7.7<br>0.03* | 140 9.5<br>0.03* | 140 9.4<br>0.03* | 137 10.8<br>0.03* | 141 10.5<br>0.03* | 88 10.0<br>0.05* |
| | | 48 7.3<br>0.04* | 46 9.3<br>0.06* | 46 9.2<br>0.04* | 52 10.7<br>0.04* | 45 10.2<br>0.06* | 26 9.6<br>0.12* |
| 5 5.7<br>0.2 | 9 5.9<br>0.2 | 11 7.3<br>0.3 | 12 7.3<br>0.5 | 14 7.9<br>0.5 | 14 11.0<br>0.6 | 12 10.1<br>0.5 | 9 9.6<br>0.4 |
| | | 6 6.9<br>0.5 | 6 6.9<br>0.5 | 9 7.5<br>0.4 | 14 10.4<br>0.5 | 11 9.7<br>0.6 | 8 9.2<br>0.4 |
| | | 9 7.89<br>0.35 | 8 8.32<br>0.45 | 6 11.02<br>0.33 | 6 10.55<br>0.36 | | |
| | | 8 7.57<br>0.05 | 8 8.17<br>0.09 | 7 10.81<br>0.25 | 7 10.22<br>0.56 | | |
| 26 5.9<br>0.3 | 35 6.2<br>0.3 | 62 7.5<br>0.4 | 74 7.7<br>0.4 | 80 8.2<br>0.5 | 84 11.3<br>0.5 | 85 10.5<br>0.4 | 71 10.1<br>0.6 |
| 14 5.6<br>0.3 | 19 6.0<br>0.3 | 25 7.1<br>0.4 | 39 7.5<br>0.4 | 53 7.9<br>0.6 | 57 10.9<br>0.5 | 61 10.1<br>0.5 | 44 9.7<br>0.7 |
| 23 6.1<br>0.4 | 32 6.6<br>0.4 | 40 8.2<br>0.6 | 40 8.3<br>0.6 | 38 8.6<br>0.6 | 32 11.5<br>0.5 | 32 11.0<br>0.5 | 26 10.4<br>0.5 |
| 6 5.7<br>0.2 | 9 6.1<br>0.3 | 11 7.5<br>0.3 | 13 7.8<br>0.5 | 14 8.0<br>0.7 | 14 11.0<br>0.3 | 13 10.4<br>0.5 | 8 10.2<br>0.7 |
| 11 6.2<br>0.4 | 10 6.7<br>0.4 | 11 8.3<br>0.4 | 10 8.3<br>0.6 | 12 8.5<br>0.5 | 13 11.2<br>0.4 | 13 10.8<br>0.6 | 9 10.4<br>0.7 |
| 4 5.6<br>0.2 | 5 6.2<br>0.6 | 5 7.6<br>0.3 | 4 8.0<br>0.4 | 4 8.4<br>0.4 | 6 10.7<br>0.5 | 6 10.4<br>0.5 | 6 9.7<br>0.5 |
| 20 6.1<br>0.2 | 21 6.5<br>0.3 | 20 8.3<br>0.4 | 22 8.2<br>0.4 | 21 8.4<br>0.3 | 20 11.1<br>0.3 | 20 10.7<br>0.4 | 14 10.1<br>0.6 |
| 17 5.8<br>0.3 | 17 6.3<br>0.4 | 17 7.4<br>0.4 | 17 7.8<br>0.6 | 16 8.2<br>0.5 | 16 10.6<br>0.4 | 13 10.4<br>0.6 | 10 9.6<br>0.9 |
| | 5 6.4<br>0.5 | 11 7.7<br>0.7 | 28 8.0<br>0.6 | 20 8.4<br>0.5 | 41 11.0<br>0.5 | 34 10.6<br>0.5 | 25 10.1<br>0.6 |
| | | 4 7.2<br>0.2 | 6 7.4<br>0.5 | 9 8.1<br>0.6 | 19 10.7<br>0.7 | 16 10.4<br>0.8 | 7 9.9<br>0.8 |
| 86 6.04<br>0.46 | 86 6.48<br>0.40 | 86 8.16<br>0.50 | 86 8.25<br>0.46 | 86 8.65<br>0.45 | 86 11.20<br>0.47 | 86 10.94<br>0.51 | |
| 96 5.84<br>0.39 | 96 6.26<br>0.41 | 96 7.47<br>0.47 | 96 7.73<br>0.51 | 96 8.33<br>0.49 | 96 10.70<br>0.44 | 96 10.42<br>0.51 | |
| | | 34 8.04<br>0.44 | 90 8.19<br>0.49 | 83 8.71<br>0.51 | 76 10.98<br>0.50 | 75 10.75<br>0.46 | |

## M e s i o d i s t a l

### MAXILLA

| | | I1 | I2 | C | P1 | P2 | M1 | M2 | M3 |
|---|---|---|---|---|---|---|---|---|---|
| Broadbeach (Smith et al., 1981) | M | 7 9.8 0.8 | 12 7.8 0.6 | 14 8.6 0.6 | 20 7.9 0.5 | 19 7.5 0.5 | 9 11.9 0.6 | 20 11.6 0.9 | 23 10.1 1.0 |
| | F | | 4 7.2 0.6 | 3 8.3 0.7 | 3 7.6 0.5 | 3 7.3 0.4 | | 4 10.9 0.6 | 3 9.7 0.4 |
| Swanport (Smith, 1982) | M | 7 9.2 0.3 | 11 7.6 0.4 | 15 8.3 0.7 | 19 7.6 0.4 | 18 7.3 0.3 | 23 11.4 0.5 | 20 10.8 0.6 | 16 9.8 0.6 |
| Wailbri (Barrett et al., 1963) | M | 130 9.35 0.58 | 115 7.65 0.63 | 80 8.31 0.57 | 98 7.69 0.46 | 96 7.19 0.43 | 115 11.34 0.52 | 82 10.70 0.71 | 36 9.87 0.82 |
| | F | 111 9.00 0.58 | 104 7.34 0.63 | 84 7.95 0.41 | 86 7.53 0.41 | 83 7.01 044 | 109 10.92 0.50 | 79 10.31 0.61 | 35 9.79 0.75 |
| Western Australia (Freedman & Lofgren, 1981) | M | 12 9.33 0.57 | 26 7.24 0.70 | 35 8.01 0.64 | 40 7.38 0.50 | 43 7.03 0.44 | 48 11.62 0.70 | 48 11.61 0.86 | 47 10.42 0.83 |
| | F | 10 8.66 0.69 | 16 7.10 0.94 | 21 7.81 0.32 | 20 7.03 0.50 | 21 6.73 0.48 | 25 10.84 0.86 | 26 10.85 0.82 | 17 9.63 0.85 |
| Yuendumu (Brown et al., 1980) | M | 106 9.42 0.50 | 103 7.54 0.58 | 99 8.27 0.52 | 101 7.64 0.45 | 99 7.18 0.43 | | | |
| | F | 66 9.16 0.55 | 63 7.36 0.64 | 61 8.01 0.40 | 62 7.54 0.39 | 60 7.10 0.35 | | | |
| Yuendumu (Barrett et al., 1963,64) | M | 130 9.35 0.05* | 115 7.65 0.06* | 80 8.31 0.06* | 98 7.69 0.05* | 96 7.19 0.04* | 115 11.34 0.05* | 82 10.70 0.08* | 36 9.87 0.14* |
| | F | 111 9.00 0.06* | 104 7.34 0.06* | 84 7.95 0.05* | 86 7.53 0.05* | 83 7.01 0.05* | 109 10.92 0.05* | 79 10.31 0.07* | 35 9.76 0.13 |
| Yuendumu (Smith, 1982) | M | 206 9.4 0.6 | 203 7.5 0.6 | 192 8.2 0.5 | 200 7.6 0.4 | 191 7.1 0.4 | 208 11.2 0.6 | 158 10.8 0.6 | 57 10.0 0.7 |
| Yuendumu (Hanihara, 1976,77) | M | 20 9.21 0.40 | 20 7.47 0.51 | 20 8.36 0.65 | 20 7.59 0.59 | 20 7.14 0.50 | 20 11.28 0.55 | 20 10.95 0.58 | |
| | F | 58 9.10 0.58 | 57 7.46 0.67 | 57 8.01 0.49 | 57 7.64 0.42 | 53 7.17 0.47 | 57 11.14 0.59 | 50 10.76 0.70 | |

## B u c c o l i n g u a l

### MAXILLA

| | | I1 | I2 | C | P1 | P2 | M1 | M2 | M3 |
|---|---|---|---|---|---|---|---|---|---|
| Broadbeach (Smith et al., 1981) | M | 8 7.9 0.6 | 14 7.0 0.6 | | 18 10.5 0.6 | 20 10.5 0.6 | 9 12.9 0.5 | 19 13.5 0.7 | 23 12.4 1.0 |
| | F | | 3 6.3 0.7 | | 3 9.7 0.4 | 3 10.2 0.2 | | 4 13.0 0.6 | 3 11.6 0.7 |
| Swanport (Smith, 1982) | M | 9 7.9 0.4 | 14 7.1 0.5 | 16 9.2 0.7 | 20 10.4 0.5 | 20 10.4 0.6 | 26 13.0 0.7 | 23 13.5 0.5 | 17 12.7 0.7 |
| Western Australia (Freedman & Lofgren, 1981) | M | 18 7.92 0.53 | 30 7.03 0.44 | 39 9.18 0.56 | 42 10.32 0.49 | 44 10.29 0.55 | 48 12.78 0.63 | 48 12.95 0.71 | 47 12.22 0.82 |
| | F | 13 7.44 0.44 | 19 6.67 0.50 | 24 8.44 0.45 | 22 9.80 0.66 | 23 9.70 0.68 | 25 12.07 0.52 | 26 12.21 0.60 | 17 11.57 0.62 |
| Yuendumu (Barrett et al., 1963,64) | M | | | 41 9.12 0.09* | 85 10.38 0.07* | 85 10.29 0.06* | 85 12.63 0.06* | 41 12.83 0.11* | 32 12.09 0.15* |
| | F | | | 36 8.67 0.06* | 81 10.10 0.06* | 81 10.05 0.07* | 81 12.21 0.06* | 36 12.42 0.11* | 30 11.93 0.15* |
| Yuendumu (Smith, 1982) | M | 205 7.9 0.5 | 194 6.9 0.5 | 186 9.0 0.6 | 199 10.2 0.5 | 191 10.2 0.5 | 208 12.5 0.6 | 154 12.8 0.7 | 48 12.3 0.8 |

MANDIBLE

| I1 | I2 | C | P1 | P2 | M1 | M2 | M3 |
|---|---|---|---|---|---|---|---|
| 8 5.9 | 15 7.0 | 17 7.7 | 16 8.1 | 18 8.0 | 7 12.3 | 16 12.8 | 19 12.1 |
| 0.3 | 0.6 | 0.4 | 0.5 | 0.6 | 0.6 | 0.6 | 0.6 |
| | | 3 7.3 | 4 7.2 | | | 4 11.9 | 4 11.3 |
| | | 0.3 | 0.9 | | | 1.0 | 0.8 |
| 136 5.87 | 130 6.60 | 98 7.49 | 95 7.49 | 89 7.56 | 119 12.04 | 82 11.45 | 33 11.61 |
| 0.40 | 0.42 | 0.46 | 0.54 | 0.51 | 0.61 | 0.68 | 0.93 |
| 117 5.68 | 112 6.36 | 95 7.01 | 85 7.36 | 82 7.31 | 101 11.62 | 80 11.07 | 33 11.32 |
| 0.43 | 0.41 | 0.38 | 0.41 | 0.44 | 0.55 | 0.65 | 0.68 |
| 11 5.55 | 14 5.86 | 22 7.28 | 23 7.15 | 25 7.17 | 27 11.62 | 28 11.75 | 30 11.72 |
| 0.63 | 0.74 | 0.51 | 0.52 | 0.46 | 0.67 | 0.80 | 0.71 |
| 12 5.04 | 14 5.77 | 15 6.91 | 18 6.81 | 15 7.01 | 15 10.55 | 15 11.25 | 17 10.91 |
| 0.40 | 0.48 | 0.44 | 0.58 | 0.62 | 0.94 | 0.61 | 1.03 |
| 106 5.88 | 106 6.60 | 102 7.46 | 100 7.49 | 95 7.63 | | | |
| 0.32 | 0.38 | 0.41 | 0.53 | 0.53 | | | |
| 66 5.74 | 66 6.46 | 64 7.04 | 61 7.35 | 60 7.42 | | | |
| 0.44 | 0.39 | 0.33 | 0.45 | 0.39 | | | |
| 136 5.87 | 130 6.60 | 98 7.49 | 95 7.49 | 89 7.56 | 119 12.04 | 82 11.45 | 33 11.61 |
| 0.03* | 0.04* | 0.05* | 0.06* | 0.05* | 0.06* | 0.08* | 0.16* |
| 117 5.68 | 112 6.36 | 95 7.01 | 85 7.36 | 82 7.31 | 101 11.62 | 80 11.07 | 33 11.32 |
| 0.04* | 0.04* | 0.04* | 0.04* | 0.05* | 0.06* | 0.07* | 0.12* |
| 20 5.74 | 20 6.43 | 20 7.40 | 20 7.40 | 20 7.62 | 20 11.97 | 20 11.53 | |
| 0.38 | 0.34 | 0.37 | 0.63 | 0.55 | 0.50 | 0.65 | |
| 58 5.67 | 57 6.42 | 58 7.03 | 58 7.46 | 51 7.45 | 58 11.68 | 43 11.29 | |
| 0.40 | 0.38 | 0.39 | 0.46 | 0.52 | 0.70 | 0.63 | |

MANDIBLE

| I1 | I2 | C | P1 | P2 | M1 | M2 | M3 |
|---|---|---|---|---|---|---|---|
| 10 6.5 | 14 6.8 | | 19 9.0 | 19 9.3 | 7 11.6 | 17 11.8 | 18 11.5 |
| 0.4 | 0.5 | | 0.4 | 0.7 | 1.0 | 0.6 | 0.5 |
| | | | 4 8.4 | 3 8.9 | | 4 11.2 | 3 10.4 |
| | | | 0.7 | 0.9 | | 0.4 | 0.5 |
| 12 6.65 | 19 6.69 | 24 8.67 | 24 8.87 | 25 9.00 | 27 11.99 | 29 11.65 | 30 11.35 |
| 0.60 | 0.48 | 0.47 | 0.65 | 0.54 | 0.69 | 0.77 | 0.72 |
| 13 5.90 | 16 6.29 | 16 7.79 | 19 8.47 | 16 8.81 | 15 11.35 | 15 11.20 | 17 10.95 |
| 0.27 | 0.45 | 0.42 | 0.50 | 0.52 | 0.73 | 0.69 | 0.93 |
| | | | 41 8.39 | 85 8.83 | 85 9.15 | 83 11.85 | 41 11.60 | 29 11.41 |
| | | | 0.08* | 0.06* | 0.07* | 0.07* | 0.10* | 0.13* |
| | | | 36 8.03 | 81 8.69 | 81 8.94 | 80 11.44 | 36 11.29 | 30 11.16 |
| | | | 0.07* | 0.06* | 0.06* | 0.06* | 0.07* | 0.10* |

## Mesiodistal

### MAXILLA

|  |  | I1 | I2 | C | P1 | P2 | M1 | M2 | M3 |
|---|---|---|---|---|---|---|---|---|---|
| Bougainville (Nasioi) (Bailit et al., 1968) | M | 62 9.00 0.67 | 75 7.75 0.61 | 79 8.57 0.51 | 82 7.54 0.51 | 90 7.03 0.51 | 85 11.10 0.62 | 61 10.13 0.74 | 33 8.97 0.93 |
|  | F | 71 8.66 0.68 | 80 7.39 0.66 | 82 8.09 0.45 | 88 7.25 0.52 | 87 6.73 0.52 | 87 10.70 0.62 | 67 9.86 0.71 | 38 8.66 0.99 |
| New Britain (West Nakanai) (Turner & Swinder, 1978) | M | 97 8.6 0.54 | 91 6.8 0.53 | 97 8.0 0.41 | 97 7.1 0.41 | 97 6.1 0.67 | 98 11.1 0.61 | 85 9.7 0.73 | 38 9.1 1.00 |
| New Guinea (Goroka) (Doran & Freedman, 1974) | M | 27 9.51 0.49 | 30 7.85 0.41 | 31 8.32 0.45 | 31 7.81 0.53 | 31 7.14 0.51 | 30 11.41 0.56 | 30 10.55 0.75 | 20 9.51 0.76 |
| New Guinea (Lufa) (Doran & Freedman, 1974) | M | 24 9.50 0.78 | 25 7.74 0.85 | 25 8.48 0.47 | 24 7.62 0.60 | 25 7.07 0.66 | 24 11.43 0.71 | 23 10.95 0.94 | 12 9.57 0.95 |

## Buccolingual

### MAXILLA

|  |  | I1 | I2 | C | P1 | P2 | M1 | M2 | M3 |
|---|---|---|---|---|---|---|---|---|---|
| Bougainville (Nasioi) (Bailit et al., 1968) | M | 52 7.80 0.53 | 63 7.06 0.68 | 73 9.15 0.85 | 74 10.29 0.83 | 83 10.34 0.67 | 82 12.26 0.73 | 63 12.46 0.77 | 39 11.66 1.03 |
|  | F | 49 7.48 0.38 | 65 6.67 0.44 | 75 8.58 0.61 | 85 9.93 0.73 | 84 9.97 0.68 | 88 11.69 0.68 | 70 11.76 0.74 | 44 10.96 1.32 |
| New Britain (West Nakanai) (Turner & Swinder, 1978) | M |  |  | 92 7.9 0.68 | 87 9.1 0.55 | 92 9.1 0.57 | 98 11.5 0.67 | 85 11.4 0.78 | 40 11.1 0.93 |
| New Guinea (Goroka) (Doran & Freedman, 1974) | M | 27 8.29 0.42 | 30 7.40 0.52 | 32 9.30 0.52 | 32 10.68 0.52 | 32 10.70 0.45 | 31 12.68 0.53 | 31 12.52 0.77 | 21 11.49 0.88 |
| New Guinea (Lufa) (Doran & Freedman, 1974) | M | 24 8.21 0.58 | 25 7.27 0.57 | 25 9.07 0.59 | 24 10.54 0.53 | 25 10.52 0.64 | 24 12.61 0.69 | 23 12.77 0.75 | 12 11.95 0.90 |

## MANDIBLE

| I1 | I2 | C | P1 | P2 | M1 | M2 | M3 |
|---|---|---|---|---|---|---|---|
| 46 5.63<br>0.39 | 68 6.35<br>0.45 | 75 7.53<br>0.45 | 82 7.58<br>0.58 | 88 7.51<br>0.62 | 87 11.80<br>0.66 | 69 11.02<br>0.71 | 39 10.92<br>0.88 |
| 68 5.57<br>0.42 | 74 6.17<br>0.42 | 77 7.03<br>0.45 | 82 7.25<br>0.54 | 84 7.12<br>0.56 | 90 11.31<br>0.72 | 74 10.61<br>0.86 | 39 10.42<br>0.77 |
| 97 5.4<br>0.41 | 97 6.0<br>0.49 | 90 7.0<br>0.42 | 92 7.0<br>0.47 | 92 7.0<br>0.51 | 98 11.5<br>0.56 | 89 10.5<br>0.74 | 57 11.0<br>0.91 |
| 27 5.79<br>0.48 | 28 6.55<br>0.42 | 29 7.49<br>0.44 | 30 7.84<br>0.57 | 30 7.77<br>0.59 | 25 12.33<br>0.65 | 28 11.53<br>0.70 | 21 11.37<br>0.99 |
| 21 6.00<br>0.31 | 23 6.48<br>0.47 | 25 7.60<br>0.41 | 21 7.76<br>0.68 | 25 7.51<br>0.65 | 22 12.10<br>0.63 | 22 11.58<br>0.57 | 18 11.16<br>0.56 |

## MANDIBLE

| I1 | I2 | C | P1 | P2 | M1 | M2 | M3 |
|---|---|---|---|---|---|---|---|
| 36 6.69<br>0.50 | 47 6.76<br>0.54 | 58 8.25<br>0.71 | 75 8.35<br>0.65 | 82 8.98<br>0.56 | 88 11.08<br>0.59 | 70 11.08<br>0.69 | 40 10.58<br>0.85 |
| 46 6.40<br>0.48 | 62 6.55<br>0.51 | 67 7.88<br>0.60 | 76 8.11<br>0.63 | 81 8.75<br>0.56 | 84 10.83<br>0.55 | 67 10.63<br>0.73 | 34 10.22<br>0.82 |
|  |  | 87 7.3<br>0.54 | 92 7.5<br>0.64 | 91 7.8<br>0.70 | 98 10.5<br>0.47 | 89 10.0<br>0.73 | 57 10.3<br>0.81 |
| 27 6.93<br>0.60 | 29 7.08<br>0.59 | 30 8.24<br>0.46 | 31 8.80<br>0.48 | 31 9.14<br>0.56 | 26 11.35<br>0.41 | 29 11.36<br>0.55 | 22 11.01<br>0.77 |
| 21 7.06<br>0.60 | 23 7.10<br>0.49 | 25 8.45<br>0.54 | 21 8.80<br>0.67 | 25 9.00<br>0.63 | 22 11.56<br>0.60 | 22 11.23<br>0.67 | 18 10.77<br>0.52 |

## Mesiodistal

**MAXILLA**

| | | I1 | I2 | C | P1 | P2 | M1 | M2 | M3 |
|---|---|---|---|---|---|---|---|---|---|
| Anglo-Saxon (Lavelle, 1968) | M | 13 8.3 0.12* | 12 6.5 0.12* | 18 8.2 0.08* | 25 6.9 0.09* | 22 6.6 0.07* | 29 9.5 0.09* | 30 9.9 0.16* | |
| | F | 13 7.8 0.15* | 12 6.1 0.07* | 18 7.8 0.05* | 25 6.8 0.19* | 22 6.3 0.19* | 29 9.3 0.15* | 30 9.5 0.12* | |
| British (Modern) (Lavelle, 1968) | M | 30 5.4 0.34* | 30 6.3 0.41* | 30 6.7 0.03* | 30 6.7 0.08* | 30 7.0 0.05* | 30 9.7 0.03* | 30 10.2 0.41* | |
| | F | 30 5.1 0.48* | 30 5.4 0.02* | 30 6.4 0.17* | 30 7.1 0.15* | 30 6.5 0.26* | 30 9.8 0.04* | 30 9.7 0.14* | |
| British (Lavelle, 1972) | M | 20 8.79 0.45 | 20 6.32 0.56 | 20 7.53 0.38 | 20 6.78 0.32 | 20 6.54 0.49 | 20 10.69 0.44 | | |
| | F | 20 8.56 0.41 | 20 6.21 0.18 | 20 7.35 0.59 | 20 6.59 0.47 | 20 6.41 0.49 | 20 10.64 0.52 | | |
| British (Contemporary) (Radnzic, 1987) | M | 60 8.97 0.39' | 60 6.87 0.21' | 60 8.04 0.14' | 60 7.09 0.18' | 60 6.91 0.31' | 60 10.69 0.31' | | |
| Iceland (Axelsson & Kirveskari, 1983) | M | 411 8.99 0.54 | 320 6.95 0.54 | 162 8.14 0.42 | 231 7.22 0.41 | 197 6.89 0.43 | 309 10.98 0.57 | 100 10.08 0.58 | |
| | F | 439 8.75 0.52 | 342 6.83 0.51 | 220 7.79 0.40 | 268 7.07 0.42 | 227 6.84 0.42 | 312 10.70 0.57 | 121 9.78 0.53 | |
| Lapps (Skolt) (Kirveskari et al., 1977) | M | 117 8.79 0.58 | 104 6.88 0.57 | 99 7.88 0.45 | 99 6.95 0.38 | 87 6.64 0.40 | 78 10.53 0.54 | 74 9.87 0.58 | 19 8.76 0.84 |
| | F | 98 8.54 0.60 | 92 6.64 0.60 | 91 7.59 0.45 | 94 6.89 0.48 | 85 6.55 0.38 | 80 10.16 0.57 | 48 9.58 0.55 | 17 8.76 0.70 |
| Norwegian Lapps (Selmer-Olsen, 1949) | M | 73 8.37 0.05* | 95 6.84 0.06* | 194 7.74 0.03* | 221 6.75 0.03* | 237 6.45 0.03* | 256 10.23 0.04* | 267 9.34 0.04* | 196 8.03 0.05* |
| | F | 85 8.34 0.05* | 105 6.70 0.05* | 177 7.47 0.03* | 205 6.55 0.03* | 208 6.32 0.03* | 223 9.93 0.03* | 225 8.93 0.04* | 154 7.85 0.06* |

## Buccolingual

**MAXILLA**

| | | I1 | I2 | C | P1 | P2 | M1 | M2 | M3 |
|---|---|---|---|---|---|---|---|---|---|
| Anglo Saxon (Lavelle, 1968) | M | 13 7.1 0.10* | 12 6.3 0.08* | 18 8.5 0.10* | 25 10.6 0.10* | 22 9.9 0.07* | 29 11.0 0.09* | 30 11.7 0.08* | |
| | F | 13 6.7 0.09* | 12 5.9 0.01* | 18 7.9 0.09* | 25 10.5 0.09* | 22 9.5 0.11* | 29 11.3 0.17* | 30 11.3 0.08* | |
| British (Modern) (Lavelle, 1968) | M | 30 5.9 0.54* | 30 6.2 0.07* | 30 7.2 0.05* | 30 8.1 0.25* | 30 9.8 0.06* | 30 11.7 0.41* | 30 12.4 0.38* | |
| | F | 30 6.1 0.03* | 30 6.1 0.25* | 30 6.8 0.09* | 30 9.4 0.05* | 30 9.0 0.03* | 30 11.8 0.41* | 30 11.5 0.26* | |
| Iceland (Axelsson & Kirveskari, 1983) | M | 234 7.35 0.49 | 127 6.58 0.49 | 115 8.58 0.59 | 212 9.43 0.57 | 202 9.60 0.63 | 351 11.92 0.52 | 89 11.77 0.77 | |
| | F | 289 7.22 0.45 | 193 6.51 0.50 | 163 8.15 0.51 | 265 9.19 0.54 | 237 9.36 0.53 | 356 11.55 0.53 | 107 11.36 0.72 | |
| Lapps (Skolt) (Kirveskari et al., 1977) | M | 89 7.44 0.43 | 75 6.57 0.49 | 82 8.46 0.47 | 99 9.30 0.49 | 96 9.11 0.51 | 88 11.73 0.52 | 80 11.66 0.63 | 18 10.81 0.92 |
| | F | 79 7.09 0.44 | 60 6.36 0.50 | 74 7.97 0.50 | 91 9.05 0.60 | 89 8.89 0.65 | 83 11.29 0.58 | 51 11.23 0.66 | 13 10.08 0.88 |
| Norwegian Lapps (Selmer-Olsen, 1949) | M | | | 197 8.18 0.04* | 203 8.90 0.04* | 208 8.72 0.04* | 231 11.24 0.04* | 258 10.96 0.04* | 187 9.93 0.06* |
| | F | | | 167 7.67 0.04* | 189 8.64 0.04* | 199 8.47 0.04* | 208 10.74 0.04* | 198 10.26 0.05* | 152 9.41 0.08* |

## MANDIBLE

| I1 | I2 | C | P1 | P2 | M1 | M2 | M3 |
|---|---|---|---|---|---|---|---|
| 14 5.2 | 18 6.9 | 17 6.8 | 25 7.2 | 26 7.0 | 29 11.3 | 30 11.6 | |
| 0.15* | 0.16* | 0.14* | 0.13* | 0.12* | 0.12* | 0.19* | |
| 14 4.1 | 18 7.1 | 17 6.6 | 25 8.0 | 26 6.1 | 29 11.2 | 30 11.4 | |
| 0.22* | 0.43* | 0.15* | 0.47* | 0.47* | 0.79* | 0.19* | |
| 14 5.4 | 20 6.1 | 30 6.7 | 30 6.9 | 30 6.7 | 30 11.8 | 30 11.9 | |
| 0.05* | 0.05* | 0.05* | 0.09* | 0.09* | 0.14* | 0.36* | |
| 14 4.7 | 20 5.9 | 30 6.5 | 30 6.6 | 30 6.9 | 30 11.7 | 30 11.9 | |
| 0.14* | 0.03* | 0.03* | 0.01* | 0.47* | 0.04* | 0.47* | |
| 20 5.58 | 20 6.20 | 20 6.96 | 20 6.79 | 20 6.81 | 20 11.36 | | |
| 0.31 | 0.44 | 0.38 | 0.55 | 0.48 | 0.66 | | |
| 20 5.56 | 20 6.17 | 20 6.91 | 20 6.77 | 20 6.78 | 20 11.21 | | |
| 0.48 | 0.31 | 0.37 | 0.42 | 0.46 | 0.39 | | |
| 60 5.56 | 60 6.15 | 60 7.25 | 60 7.32 | 60 7.35 | 60 11.23 | | |
| 0.12' | 0.13' | 0.16' | 0.16' | 0.19' | 0.47' | | |
| 465 5.59 | 431 6.20 | 217 7.13 | 225 7.30 | 200 7.45 | 280 11.45 | 115 10.85 | |
| 0.35 | 0.36 | 0.41 | 0.41 | 0.46 | 0.58 | 0.60 | |
| 482 5.48 | 442 6.02 | 307 6.80 | 309 7.12 | 239 7.27 | 287 11.12 | 158 10.49 | |
| 0.34 | 0.37 | 0.35 | 0.42 | 0.44 | 0.60 | 0.64 | |
| 129 5.50 | 140 6.07 | 144 7.02 | 127 6.97 | 112 7.01 | 57 11.26 | 66 10.85 | 17 10.53 |
| 0.32 | 0.33 | 0.36 | 0.38 | 0.35 | 0.55 | 0.63 | 1.23 |
| 119 5.37 | 133 5.88 | 129 6.63 | 111 6.87 | 97 6.88 | 61 10.95 | 60 10.56 | 13 10.31 |
| 0.34 | 0.37 | 0.34 | 0.40 | 0.45 | 0.66 | 0.72 | 0.83 |
| 76 5.36 | 123 5.98 | 219 6.82 | 226 6.72 | 232 6.74 | 228 10.95 | 254 10.51 | 172 9.90 |
| 0.03* | 0.04* | 0.03* | 0.03* | 0.03* | 0.04* | 0.04* | 0.07* |
| 83 5.22 | 124 5.85 | 186 6.50 | 191 6.59 | 191 6.59 | 192 10.64 | 203 10.06 | 152 9.57 |
| 0.03* | 0.04* | 0.03* | 0.03* | 0.03* | 0.04* | 0.04* | 0.06* |

## MANDIBLE

| I1 | I2 | C | P1 | P2 | M1 | M2 | M3 |
|---|---|---|---|---|---|---|---|
| 14 6.0 | 18 6.1 | 17 6.9 | 25 6.7 | 26 8.0 | 29 9.7 | 30 10.2 | |
| 0.38* | 0.42* | 0.46* | 0.13* | 0.12* | 0.17* | 0.19* | |
| 14 5.2 | 18 6.2 | 17 7.2 | 25 8.3 | 26 8.8 | 29 9.4 | 30 9.8 | |
| 0.43* | 0.44* | 0.50* | 0.58* | 0.62* | 0.65* | 0.65* | |
| 14 5.9 | 20 6.2 | 30 7.1 | 30 8.6 | 30 9.2 | 30 9.9 | 30 10.8 | |
| 0.09* | 0.14* | 0.47* | 0.01* | 0.01* | 0.02* | 0.23* | |
| 14 5.7 | 20 5.7 | 30 6.8 | 30 8.2 | 30 7.3 | 30 9.8 | 30 10.2 | |
| 0.21* | 0.07* | 0.05* | 0.01* | 0.01* | 0.49* | 0.52* | |
| 194 6.07 | 143 6.42 | 95 7.76 | 177 8.00 | 161 8.72 | 279 10.91 | 87 10.67 | |
| 0.42 | 0.40 | 0.54 | 0.50 | 0.61 | 0.46 | 0.46 | |
| 296 5.94 | 250 6.34 | 194 7.41 | 240 7.71 | 201 8.44 | 276 10.64 | 130 10.29 | |
| 0.37 | 0.38 | 0.47 | 0.50 | 0.57 | 0.45 | 0.52 | |
| 99 6.18 | 96 6.55 | 111 7.96 | 111 7.61 | 105 8.10 | 61 10.94 | 68 10.62 | 13 10.26 |
| 0.37 | 0.38 | 0.39 | 0.43 | 0.53 | 0.47 | 0.54 | 0.47 |
| 102 5.88 | 100 6.32 | 106 7.36 | 97 7.42 | 93 7.88 | 63 10.68 | 57 10.35 | 13 10.09 |
| 0.39 | 0.39 | 0.42 | 0.50 | 0.66 | 0.50 | 0.56 | 0.69 |
| | | 211 7.55 | 226 7.39 | 217 7.75 | 246 10.40 | 259 10.07 | 168 9.52 |
| | | 0.03* | 0.03* | 0.03* | 0.04* | 0.04* | 0.06* |
| | | 190 6.95 | 197 7.13 | 203 7.51 | 207 10.50 | 218 9.62 | 152 9.20 |
| | | 0.03* | 0.03* | 0.03* | 0.04* | 0.04* | 0.06* |

M e s i o d i s t a l

**MAXILLA**

| | | I1 | I2 | C | P1 | P2 | M1 | M2 | M3 |
|---|---|---|---|---|---|---|---|---|---|
| American (Hanihara & Ueda, 1979) | M | 38  8.64  0.61 | 38  6.56  0.57 | 37  7.81  0.46 | 38  7.09  0.40 | 38  6.78  0.51 | 38  10.72  0.64 | | |
| American (Hanihara, 1976) | M | 47  8.63  0.65 | 46  6.63  0.59 | 45  7.82  0.45 | 45  7.08  0.44 | 46  6.75  0.51 | 47  10.73  0.60 | 40  10.25  0.60 | |
| | F | 33  8.41  0.52 | 32  6.51  0.56 | 34  7.57  0.35 | 34  6.85  0.40 | 32  6.62  0.46 | 34  10.38  0.45 | 24  9.67  0.44 | |
| American (Takehisa, 1957) | M | 102  8.57  0.63 | 96  6.68  0.67 | 116  7.86  0.47 | 103  7.00  0.58 | 86  6.88  0.61 | 73  10.48  0.81 | 85  9.85  1.04 | |
| American (Garn et al., 1967b) | M | 208  8.78  0.57 | 198  6.71  0.58 | 183  7.95  0.45 | 197  7.14  0.47 | 193  6.84  0.46 | 207  10.17  0.49 | 150  10.05  0.58 | |
| | F | 264  8.50  0.57 | 253  6.47  0.67 | 239  7.51  0.44 | 254  6.90  0.41 | 234  6.60  0.60 | 262  9.81  0.40 | 201  9.63  0.61 | |
| American (Moorrees et al., 1957) | M | 87  8.78  0.46 | 84  6.64  0.63 | 87  7.95  0.42 | 87  7.01  0.38 | 86  6.82  0.37 | 83  10.81  0.56 | 65  10.35  0.63 | |
| | F | 87  8.40  0.53 | 86  6.47  0.62 | 85  7.53  0.37 | 84  6.85  0.42 | 81  6.62  0.43 | 85  10.52  0.51 | 50  9.81  0.49 | |
| South Africa (Kieser et al., 1985e) | M | 57  8.94  0.70 | 55  7.08  0.54 | 55  8.43  0.59 | 54  7.53  0.51 | 55  7.49  0.63 | 54  11.22  0.65 | 55  10.71  0.67 | |
| | F | 66  8.40  0.66 | 66  6.56  0.57 | 66  7.74  0.42 | 61  7.24  0.45 | 66  7.04  0.41 | 66  10.74  0.50 | 66  10.00  0.49 | |
| Tristan da Cunha (Thomsen, 1955; quoted by Jacobson, 1982) | M | 152  8.78  0.5* | 146  6.74  0.05* | 132  7.93  0.04* | 132  6.96  0.04* | 128  6.64  0.04* | 145  10.69  0.05* | 109  10.03  0.08* | 63  8.89  0.10* |
| | F | 131  8.60  0.05* | 131  6.68  0.07* | 112  7.74  0.04* | 112  7.07  0.07* | 102  6.59  0.03* | 122  10.45  0.04* | 96  9.78  0.07* | 56  8.87  0.10* |
| Australian (Townsend & Alvesalo, 1985) | M | 104  8.76  0.53 | 95  6.85  0.55 | 82  8.04  0.43 | 74  7.07  0.39 | 74  6.77  0.44 | 81  10.65  0.50 | 65  10.31  0.57 | |

B u c c o l i n g u a l

**MAXILLA**

| | | I1 | I2 | C | P1 | P2 | M1 | M2 | M3 |
|---|---|---|---|---|---|---|---|---|---|
| American (Takehisa, 1957) | M | 101  7.02  0.61 | 96  6.28  0.67 | 116  8.15  0.66 | 100  8.46  0.36 | 87  9.05  0.27 | 73  11.44  0.59 | 85  11.38  0.62 | |
| South Africa (Kieser et al., 1985e) | M | | | | 52  9.25  0.51 | 55  9.37  0.61 | 55  11.53  0.60 | 46  11.17  0.62 | |
| | F | | | | 61  8.68  0.62 | 66  8.98  0.52 | 66  10.95  0.60 | 53  10.55  0.63 | |
| Tristan da Cunha (Thomsen, 1955; quoted by Jacobson, 1982) | M | | | 82  9.38  0.06* | 75  9.81  0.06* | 65  9.96  0.08* | 104  12.14  0.06* | 84  12.33  0.09* | 29  11.82  0.21* |
| | F | | | 47  8.87  0.08* | 49  9.64  0.09* | 41  9.83  0.07* | 70  11.77  0.06* | 47  11.97  0.09* | 23  12.12  0.20* |
| Australian (Townsend & Alvesalo, 1985) | M | 81  7.34  0.52 | 71  6.57  0.57 | 81  8.62  0.55 | 76  9.47  0.56 | 83  9.50  0.60 | 98  11.76  0.55 | 69  11.81  0.72 | |

## MANDIBLE

| I1 | I2 | C | P1 | P2 | M1 | M2 | M3 |
|---|---|---|---|---|---|---|---|
| 38 5.46 / 0.42 | 38 5.97 / 0.45 | 38 6.85 / 0.42 | 38 7.09 / 0.41 | 38 7.30 / 0.49 | 38 11.22 / 0.68 | | |
| 43 5.50 / 0.43 | 44 6.04 / 0.47 | 45 6.93 / 0.45 | 44 7.10 / 0.39 | 41 7.30 / 0.50 | 45 11.24 / 0.65 | 34 10.83 / 0.72 | |
| 30 5.22 / 0.31 | 30 5.78 / 0.39 | 30 6.48 / 0.38 | 30 7.00 / 0.36 | 26 7.07 / 0.51 | 30 10.74 / 0.54 | 22 10.25 / 0.55 | |
| 144 5.51 / 0.48 | 147 5.10 / 0.46 | 161 6.95 / 0.53 | 145 6.96 / 0.48 | 101 7.47 / 0.56 | 37 10.94 / 0.66 | 75 11.00 / 1.10 | |
| 204 5.38 / 0.38 | 212 6.02 / 0.42 | 203 6.98 / 0.40 | 200 7.27 / 0.47 | 177 7.26 / 0.46 | 103 11.38 / 0.55 | 128 10.63 / 0.62 | |
| 258 5.31 / 0.39 | 262 5.86 / 0.40 | 258 6.56 / 0.41 | 254 6.99 / 0.40 | 237 7.03 / 0.40 | 258 10.86 / 0.67 | 166 10.18 / 0.60 | |
| 85 5.42 / 0.31 | 85 5.95 / 0.38 | 84 6.96 / 0.36 | 85 7.07 / 0.35 | 82 7.29 / 0.52 | 76 11.18 / 0.47 | 53 10.76 / 0.71 | |
| 87 5.25 / 0.36 | 87 5.78 / 0.38 | 87 6.47 / 0.32 | 87 6.87 / 0.38 | 83 7.02 / 0.40 | 84 10.74 / 0.56 | 53 10.34 / 0.62 | |
| 55 5.54 / 0.32 | 55 6.20 / 0.43 | 55 7.34 / 0.48 | 54 7.68 / 0.50 | 52 7.81 / 0.51 | 55 11.56 / 0.58 | 37 10.80 / 0.62 | |
| 65 5.33 / 0.37 | 65 6.01 / 0.46 | 65 6.79 / 0.36 | 62 7.30 / 0.53 | 60 7.38 / 0.44 | 65 10.88 / 0.55 | 52 10.20 / 0.59 | |
| 154 5.54 / 0.04* | 152 6.08 / 0.05* | 136 7.15 / 0.04* | 135 7.07 / 0.06* | 121 7.21 / 0.05* | 143 11.22 / 0.05* | 108 10.77 / 0.07* | 45 10.78 / 0.14* |
| 132 5.49 / 0.04* | 135 6.08 / 0.04* | 120 6.87 / 0.04* | 120 7.10 / 0.05* | 107 7.13 / 0.04* | 112 11.01 / 0.05* | 92 10.51 / 0.07* | 43 10.40 / 0.13* |
| 100 5.47 / 0.32 | 102 6.04 / 0.35 | 87 6.99 / 0.40 | 76 7.19 / 0.46 | 79 7.29 / 0.46 | 72 11.42 / 0.59 | 54 10.94 / 0.65 | |

## MANDIBLE

| I1 | I2 | C | P1 | P2 | M1 | M2 | M3 |
|---|---|---|---|---|---|---|---|
| 144 5.91 / 0.70 | 147 6.18 / 0.78 | 161 7.57 / 0.65 | 144 7.76 / 0.63 | 102 8.43 / 0.58 | 37 10.40 / 0.54 | 75 10.27 / 0.61 | |
| | | 53 8.18 / 0.68 | 53 8.68 / 0.62 | 55 10.34 / 0.53 | 45 10.43 / 0.59 | | |
| | | 64 7.67 / 0.44 | 63 8.11 / 0.52 | 65 10.34 / 0.53 | 52 9.95 / 0.56 | | |
| | | 54 8.97 / 0.09* | 64 8.81 / 0.06* | 83 9.21 / 0.05* | 110 11.22 / 0.05* | 80 11.16 / 0.07* | 14 11.53 / 0.16* |
| | | 43 8.45 / 0.10* | 61 8.69 / 0.07* | 55 9.16 / 0.07* | 100 11.08 / 0.07* | 72 10.94 / 0.07* | 13 11.05 / 0.17* |
| 78 6.18 / 0.45 | 78 6.44 / 0.42 | 73 7.92 / 0.53 | 74 8.13 / 0.54 | 78 8.64 / 0.53 | 92 10.83 / 0.49 | 67 10.72 / 0.60 | |

M e s i o d i s t a l

| | | I1 | I2 | C | P1 | P2 | M1 | M2 | M3 |
|---|---|---|---|---|---|---|---|---|---|
| **MAXILLA** | | | | | | | | | |
| Bedouin (Israeli) (Rosenzweig & Silberman, 1969) | M | 83 8.76 0.55 | 80 6.89 0.48 | 58 7.80 0.49 | 62 7.02 0.46 | 60 6.75 0.47 | 78 10.63 0.58 | 35 10.03 0.70 | |
| | F | 54 8.49 0.59 | 53 6.79 0.63 | 46 7.46 0.49 | 45 6.72 0.39 | 46 6.53 0.59 | 53 10.22 0.62 | 38 9.52 0.57 | |
| Druse (Israel) (Koyoumdjisky-Kaye et al., 1977) | M | 45 8.58 0.57 | 46 6.46 0.53 | 34 7.79 0.43 | 44 6.82 0.40 | 44 6.55 0.42 | 45 10.72 0.68 | 13 9.86 0.63 | |
| | F | 39 8.41 0.58 | 42 6.36 0.68 | 26 7.53 0.43 | 37 6.80 0.34 | 32 6.46 0.30 | 40 10.13 0.65 | | |
| Jewish Cochini (Rosenzweig & Ziberman, 1967) | M | 30 8.44 0.73 | 30 6.61 0.36 | 29 7.69 0.44 | 30 7.04 0.42 | 28 6.73 0.32 | 28 10.56 0.53 | 21 9.66 0.48 | |
| | F | | | | 30 9.07 0.43 | 28 8.83 0.71 | 28 11.05 0.63 | 21 10.82 0.77 | |
| Jewish (Indian extraction) (Koyoumdjisky-Kaye et al., 1976) | M | 51 8.43 0.50 | 51 6.67 0.49 | 32 7.76 0.38 | 34 7.05 0.45 | 28 6.64 0.38 | 48 10.57 0.47 | 5 10.41 0.45 | |
| | F | 38 8.15 0.56 | 38 6.58 0.52 | 30 7.36 0.34 | 30 6.86 0.45 | 28 6.59 0.47 | 38 10.30 0.59 | | |
| Jewish (Kurdish extraction) (Koyoumdjisky-Kaye et al., 1976) | M | 49 8.79 0.66 | 49 6.58 0.73 | 27 7.81 0.49 | 43 6.88 0.49 | 36 6.65 0.43 | 48 10.16 0.57 | 14 8.92 0.59 | |
| | F | 50 8.38 0.47 | 49 6.36 0.53 | 43 7.42 0.50 | 48 6.59 0.36 | 40 6.34 0.47 | 50 9.81 0.54 | 28 8.77 0.85 | |
| Jewish (Yeminite) (Koyoumdjisky-Kaye et al., 1976) | M | 48 8.49 0.65 | 48 6.44 0.55 | 20 7.43 0.57 | 42 6.82 0.41 | 36 6.52 0.56 | 47 10.32 0.71 | | |
| | F | 59 8.39 0.54 | 59 6.32 0.74 | 48 7.43 0.47 | 51 6.77 0.44 | 49 6.49 0.48 | 59 10.38 0.56 | 14 9.77 0.76 | |
| Jewish (Yemenite) Rosenzweig & Zilberman, 1967 | M | 30 8.06 0.54 | 30 6.32 0.37 | 20 7.51 0.66 | 28 6.88 0.41 | 25 6.49 0.64 | 30 10.12 0.63 | 16 9.21 0.66 | |
| | F | 30 8.42 0.69 | 26 6.26 0.85 | 20 7.28 0.56 | 29 6.78 0.55 | 24 6.61 0.52 | 30 10.44 0.62 | 15 9.72 0.81 | |
| Circassian (Israel) (Koyoumdjisky-Kaye et al., 1977) | M | 57 8.72 0.54 | 59 6.65 0.47 | 28 7.80 0.36 | 55 6.99 0.45 | 43 6.71 0.39 | 59 10.74 0.54 | | |
| | F | 59 8.39 0.54 | 59 6.32 0.74 | 48 7.43 0.47 | 51 6.77 0.44 | 49 6.49 0.48 | 59 10.38 0.56 | 14 9.77 0.76 | |
| Iraqi (Ghose & Baghdady, 1979) | M | 78 9.03 0.64 | 78 6.95 0.75 | 78 8.06 0.61 | 78 7.17 0.53 | 78 6.94 0.50 | 78 10.70 0.60 | | |
| | F | 83 8.84 0.60 | 83 6.87 0.69 | 83 7.84 0.53 | 83 7.06 0.55 | 83 6.92 0.53 | 83 10.62 0.67 | | |

## MANDIBLE

| I1 | I2 | C | P1 | P2 | M1 | M2 | M3 |
|---|---|---|---|---|---|---|---|
| 75 5.45<br>0.37 | 77 6.09<br>0.39 | 59 7.03<br>0.49 | 63 7.03<br>0.46 | 51 7.21<br>0.47 | 75 11.27<br>0.55 | 39 10.49<br>0.69 | |
| 51 5.23<br>0.33 | 53 5.88<br>0.44 | 47 6.52<br>0.44 | 47 6.78<br>0.40 | 46 6.92<br>0.44 | 54 10.76<br>0.64 | 40 10.05<br>0.50 | |
| 46 5.28<br>0.31 | 46 5.78<br>0.40 | 39 6.82<br>0.50 | 40 6.91<br>0.43 | 35 7.01<br>0.48 | 42 11.07<br>0.60 | 34 10.30<br>0.59 | |
| 41 5.20<br>0.37 | 40 5.76<br>0.44 | 37 6.49<br>0.46 | 35 6.70<br>0.31 | 30 6.78<br>0.34 | 40 10.64<br>0.60 | 19 9.86<br>0.68 | |
| 28 5.40<br>0.21 | 28 5.81<br>0.40 | 30 6.79<br>0.51 | 28 6.90<br>0.54 | 29 7.00<br>0.38 | 29 11.15<br>0.57 | 22 9.93<br>0.56 | |
| | | | 28 7.45<br>0.63 | 29 7.97<br>0.35 | 29 10.43<br>0.60 | 22 9.80<br>0.43 | |
| 47 5.25<br>0.30 | 50 5.82<br>0.40 | 43 6.71<br>0.42 | 41 7.17<br>0.44 | 30 7.20<br>0.42 | 49 11.17<br>0.52 | 8 10.40<br>0.66 | |
| 36 5.15<br>0.39 | 36 5.72<br>0.49 | 36 6.32<br>0.33 | 31 6.90<br>0.80 | 28 7.15<br>0.48 | 35 10.75<br>0.44 | 7 10.46<br>0.79 | |
| 49 5.42<br>0.54 | 49 5.93<br>0.50 | 40 6.92<br>0.54 | 41 6.91<br>0.51 | 31 7.07<br>0.58 | 45 10.95<br>0.79 | 14 10.05<br>0.83 | |
| 49 5.36<br>0.44 | 49 5.84<br>0.45 | 46 6.46<br>0.43 | 44 6.64<br>0.40 | 38 6.87<br>0.54 | 50 10.53<br>0.52 | 30 9.73<br>0.61 | |
| 48 5.31<br>0.38 | 48 5.69<br>0.45 | 34 6.49<br>0.55 | 33 6.73<br>0.48 | 22 7.06<br>0.57 | 47 10.74<br>0.77 | 11 9.97<br>0.50 | |
| 58 5.29<br>0.37 | 56 5.88<br>0.48 | 52 6.49<br>0.41 | 52 6.74·<br>0.43 | 46 6.88<br>0.43 | 57 10.93<br>0.51 | 24 10.14<br>0.59 | |
| 30 6.17<br>0.28 | 30 5.61<br>0.41 | 27 6.49<br>0.24 | 28 6.70<br>0.45 | 25 6.97<br>0.64 | 30 10.83<br>0.75 | 23 9.86<br>0.65 | |
| 30 5.23<br>0.43 | 30 5.62<br>0.48 | 27 6.43<br>0.40 | 27 6.59<br>0.64 | 17 6.98<br>0.48 | 30 10.75<br>0.34 | 18 9.82<br>0.63 | |
| 58 5.47<br>0.34 | 59 6.00<br>0.40 | 40 6.88<br>0.46 | 45 6.93<br>0.46 | 29 7.15<br>0.40 | 55 11.21<br>0.54 | 12 10.14<br>0.40 | |
| 58 5.29<br>0.37 | 56 5.88<br>0.48 | 52 6.49<br>0.41 | 52 6.74<br>0.43 | 46 6.88<br>0.43 | 57 10.93<br>0.51 | 24 10.14<br>0.59 | |
| 78 5.61<br>0.44 | 78 6.20<br>0.50 | 78 6.97<br>0.44 | 78 7.12<br>0.48 | 78 7.36<br>0.57 | 78 11.26<br>0.69 | | |
| 83 5.66<br>0.45 | 83 6.19<br>0.45 | 83 6.78<br>0.52 | 83 7.04<br>0.61 | 83 7.28<br>0.52 | 83 11.03<br>0.64 | | |

## Buccolingual

### MAXILLA

| | | I1 | I2 | C | P1 | P2 | M1 | M2 | M3 |
|---|---|---|---|---|---|---|---|---|---|
| Bedouin (Israeli) | M | | | | 62 9.20 0.66 | 60 9.42 0.63 | 78 11.54 0.65 | 35 11.55 0.76 | |
| (Rosenzweig & Silberman, 1969) | F | | | | 45 6.72 0.39 | 46 6.53 0.59 | 53 10.22 0.62 | 38 9.52 0.57 | |
| Druse (Israel) | M | | | 29 7.82 0.67 | 45 8.92 0.54 | 44 9.25 0.56 | 46 11.40 0.61 | 27 11.01 0.63 | |
| (Koyoumdjisky-Kaye et al., 1977) | F | | | 21 7.62 0.57 | 39 8.68 0.47 | 32 8.88 0.42 | 42 10.98 0.59 | 5 10.82 0.44 | |
| Jewish Cochini | M | 30 8.05 0.29 | 30 6.34 0.54 | 30 7.22 0.23 | 29 6.64 0.42 | 30 6.33 0.45 | 30 10.10 0.51 | 26 9.37 0.62 | |
| (Rosenzweig & Ziberman, 1967) | F | | | | 29 8.48 0.40 | 30 8.31 0.56 | 30 10.41 0.32 | 26 10.05 0.46 | |
| Jewish (Indian extraction) | M | | | 25 8.11 0.56 | 33 9.40 0.58 | 29 9.40 0.61 | 49 11.46 0.61 | 11 11.40 0.84 | |
| (Koyoumdjisky-Kaye et al., 1976) | F | | | 26 7.66 0.55 | 31 9.01 0.51 | 29 9.04 0.53 | 38 10.94 0.47 | 5 10.82 0.53 | |
| Jewish (Kurdish extraction) | M | | | 29 7.29 0.92 | 44 8.91 0.52 | 37 9.08 0.52 | 49 10.87 0.63 | 19 10.22 0.72 | |
| (Koyoumdjisky-Kaye et al., 1976) | F | | | 35 7.43 0.56 | 51 8.72 0.62 | 50 9.00 0.70 | 58 11.05 0.60 | 20 10.39 0.57 | |
| Jewish (Yeminite) | M | | | 15 7.49 0.95 | 42 8.70 0.59 | 33 8.90 0.71 | 47 10.81 0.63 | 13 10.55 0.70 | |
| (Koyoumdjisky-Kaye et al., 1976) | F | | | 35 7.43 0.56 | 51 8.72 0.62 | 50 9.00 0.70 | 58 11.05 0.60 | 20 10.39 0.57 | |
| Jewish (Yemenite) | M | | | | 28 8.69 0.65 | 25 8.83 0.61 | 30 10.95 0.69 | 16 10.34 0.65 | |
| Rosenzweig & Zilberman, 1967 | F | | | | 29 8.39 0.63 | 24 8.64 0.79 | 30 10.86 0.72 | 15 10.51 0.92 | |
| Circassian (Israel) | M | | | 20 7.65 0.51 | 55 9.06 0.56 | 45 9.27 0.57 | 59 11.38 0.62 | 11 10.66 0.82 | |
| (Koyoumdjisky-Kaye et al., 1977) | F | | | 35 7.43 0.56 | 51 8.72 0.62 | 50 9.00 0.60 | 58 10.05 0.60 | 20 10.39 0.57 | |

MANDIBLE

| I1 | I2 | C | P1 | P2 | M1 | M2 | M3 |
|---|---|---|---|---|---|---|---|
| | | | 63 7.96 0.48 | 51 8.49 0.71 | 75 10.77 0.53 | 39 10.44 0.61 | |
| | | | 47 7.40 0.49 | 46 8.04 0.54 | 54 10.34 0.49 | 40 10.06 0.52 | |
| | | 30 6.96 0.60 | 39 7.47 0.52 | 35 8.02 0.48 | 42 10.37 0.49 | 36 9.92 0.49 | |
| | | 28 6.80 0.66 | 34 7.15 0.45 | 30 7.86 0.43 | 41 10.11 0.50 | 19 9.50 0.40 | |
| 30 5.08 0.36 | 30 5.68 0.37 | 29 6.36 0.38 | 30 6.63 0.53 | 28 6.80 0.51 | 29 10.63 0.58 | 24 9.43 0.58 | |
| | | | 30 7.05 0.51 | 28 7.43 0.54 | 29 9.80 0.51 | 24 9.13 0.46 | |
| | | 24 7.19 0.74 | 36 8.05 0.64 | 29 8.64 0.55 | 50 10.67 0.60 | 12 10.49 0.70 | |
| | | 30 6.87 0.50 | 30 7.70 0.57 | 28 8.32 0.61 | 36 10.37 0.50 | 10 10.02 0.79 | |
| | | 40 6.02 1.04 | 30 7.31 0.52 | 48 8.19 0.62 | 26 10.19 0.66 | 41 9.53 0.81 | |
| | | 42 6.79 0.53 | 51 7.37 0.56 | 44 8.16 0.57 | 57 10.37 0.59 | 28 9.74 0.60 | |
| | | 27 6.31 0.98 | 33 7.15 0.56 | 21 8.00 0.68 | 48 10.03 0.59 | 20 9.54 0.71 | |
| | | 42 6.79 0.53 | 51 7.37 0.56 | 44 8.16 0.57 | 57 10.37 0.59 | 28 9.74 0.60 | |
| | | | 28 7.25 0.50 | 25 7.90 0.55 | 30 10.20 0.36 | 23 9.53 0.58 | |
| | | | 27 6.96 0.55 | 17 7.92 0.45 | 30 10.12 0.41 | 18 9.71 0.44 | |
| | | 28 6.82 0.64 | 44 7.46 0.51 | 30 8.26 0.60 | 59 10.65 0.50 | 17 10.01 0.78 | |
| | | 42 6.79 0.53 | 51 7.37 0.56 | 44 8.16 0.57 | 57 10.37 0.59 | 28 9.74 0.60 | |

# References

Adams, M. & Niswander, J. D. (1967). Developmental noise and a congenital malformation. *Genetic Research (Cambridge)*, **10**, 313–17.

Aiello, L. C. (1981). The allometry of primate body proportions. *Symposia of the Zoological Society of London*, **48**, 331–58.

Alberch, P. (1980). Ontogenesis and morphological diversification. *American Zoologist*, **20**, 653–67.

Alberch, P. (1982). Developmental constraints in evolutionary process. In *Evolution and Development*, ed. J. T. Bonner, pp. 313–32. Berlin: Springer.

Alberch, P. & Alberch, J. (1982). Heterochronic mechanisms of morphological diversification and evolutionary change in the neotropical salamander *Bolitoglossa occidentalis. Journal of Morphology*, **167**, 249–64.

Albrecht, G. A. (1978). Craniofacial morphology of Sulawesi macaques. *Contributions to Primatology*, **13**, 1–151.

Albrecht, G. A. (1980). Multivariate analysis and the study of form, with special reference to canonical variate analysis. *American Zoologist*, **20**, 679–93.

Alexander, R. McN. (1982). Size, shape and structure for running and flight. In *A Companion to Animal Physiology*, ed. C. R. Taylor, K. Johansen & L. Bolis, pp. 309–24. Cambridge: Cambridge University Press.

Alvesalo, L. (1971). The influence of sex chromosome genes on tooth size in man. *Suom Hammaslaak Toim*, **67**, 3–54.

Alvesalo, L. & Kari, M. (1977). Sizes of deciduous teeth in 47 XYY males. *American Journal of Human Genetics*, **29**, 486–9.

Alvesalo, L., Osborne, R. H. & Kari, M. (1975). The 47 XYY males' Y chromosome and tooth size. *American Journal of Human Genetics*, **27**, 53–61.

Alvesalo, L. & Tigerstedt, P. M. A. (1974). Heritabilities of human tooth dimensions. *Hereditas, 77*, 311–18.

Amadon, D. (1975). Why are female birds larger than males? *Raptor Research*, **9**, 1–11.

Anderson, D. L. & Thompson, G. W. (1973). Interrelationships and sex differences of dental and skeletal measurements. *Journal of Dental Research*, **52**, 431–8.

Anderson, D. L., Thompson, G. W. & Popovich, F. (1977). Tooth, chin, bone and body size correlations. *American Journal of Physical Anthropology*, **46**, 7–12.

Anderson, T. W. (1958). *An Introduction to Multivariate Statistical Analysis*. New York: John Wiley.

Arya, B. S., Savara, B. S., Thomas, D. & Clarkson, D. (1974). Relation of sex and occlusion to mesiodistal tooth size. *American Journal of Orthodontics*, **66**, 479–86.

Ashley-Montagu, M. F. (1940). The significance of the variability of the upper lateral incisor teeth in man. *Human Biology*, **12**, 323–59.

Atchley, W. R. (1983). Some genetic aspects of morphometric variation. In *Numerical Taxonomy*, ed. J. Felsenstein, pp. 346–63. Berlin: Springer.

Atchley, W. R. (1984). Ontogeny, timing of development and genetic variance – covariance structure. *American Naturalist*, **123**, 519–40.

Atchley, W. R. & Rutledge, J. J. (1980). Genetic components of size and shape. I. Dynamics of components of phenotypic variability and covariability during ontogeny in the laboratory rat. *Evolution*, **34**, 1161–73.

Axellson, G. & Kirveskari, P. (1983). Crown size of permanent teeth in Icelanders. *Acta Odontologica Scandinavica*, **41**, 181–6.

Bachrach, H. & Young, M. (1927). A comparison of the degree of resemblance in dental characters shown in pairs of twins of identical and fraternal types. *British Dental Journal*, **21**, 1293–304.

Bader, R. S. (1965). Fluctuating asymmetry in the dentition of the house mouse. *Growth*, **29**, 291–300.

Bailey, D. W. (1956). A comparison of genetic and environmental influences on the shape of the axis in mice. *Genetics*, **41**, 207–22.

Bailit, H. L., Damon, S. & Damon, A. (1966). Consanguinity in Tristan da Cunha in 1938. *Eugenics Quarterly*, **13**, 30–3.

Bailit, H. L., de Witt, S. J. & Leigh, R. A. (1968). The size and morphology of the Nasioi dentition. *American Journal of Physical Anthropology*, **28**, 271–88.

Bailit, H. L. & Friedlaender, J. S. (1966). Tooth size reduction: a hominid trend. *American Anthropologist*, **68**, 665–72.

Bailit, H. L. & Sung, B. (1968). Maternal effects on the developing dentition. *Archives of Oral Biology*, **13**, 155–61.

Bailit, H. L., Workman, P. L., Niswander, J. D. & MacLean, C. J. (1970). Dental asymmetry as an indicator of genetic and environmental conditions in human populations. *Human Biology*, **42**, 626–38.

Barden, H. S. (1980). Fluctuating dental asymmetry: a measure of developmental instability in Down's syndrome. *American Journal of Physical Anthropology*, **52**, 169–73.

Barnes, D. S. (1969). Tooth morphology and other aspects of the Teso Dentition. *American Journal of Physical Anthropology*, **30**, 183–94.

Barrett, M. J., Brown, T. & MacDonald, M. R. (1963). Dental observations on Australian Aborigines: MD crown diameters of permanent teeth. *Australian Dental Journal*, **8**, 150–6.

Barrett, M. J., Brown, T., Arato, G. & Ozols, I. V. (1964). Dental observations on Australian Aborigines: BL diameters of deciduous and permanent teeth. *Australian Dental Journal*, **9**, 280–5.

Bateson, W. (1894). *Materials for the Study of Variation, Treated with Special Regard to Discontinuity in the Origin of Species*. London: Methuen.

Bateson, W. (1913). *Problems of Genetics*. New Haven: Yale University Press.

Beardmore, J. A. (1960). Developmental stability in constant and fluctuating temperatures. *Heredity*, **14**, 411–22.

Beresford, J. S. (1969). Tooth size and class distinction. *Dental Practitioner*, **20**, 113–20.

Bhasin, M. K., Sharma, A., Singh, I. P. & Walter, H. (1985). Morphological and metrical dental study on Indians. *Zeitschrift für Morphologie und Anthropologie*, **76**, 77–90.

Biggerstaff, R. H. (1970). Morphological variation for the permanent mandibular first molars in human monozygotic and dizygotic twins. *Archives of Oral Biology*, **15**, 721–33.

Biggerstaff, R. H. (1979). The biology of dental genetics. *Yearbook of Physical Anthropology*, **22**, 215–27.

Bjork, A. (1947). *The Face in Profile*. Lund: Berlinska.

Black, T. K. (1978). Sexual dimorphism in the tooth crown diameters of the deciduous teeth. *American Journal of Physical Anthropology*, **48**, 77–82.

Black, T. K. (1980). An exception to the apparent relationships between stress and fluctuating dental asymmetry. *Journal of Dental Research*, **59**, 1168–9.

Bland, J. M. & Altman, D. G. (1986). Statistical methods for assessing agreement between two methods of clinical measurement. *Lancet*, **1**, 307–10.

Boklage, C. E. (1987). Developmental differences between singletons and twins in distributions of dental diameter asymmetries. *American Journal of Physical Anthropology*, **74**, 319–31.

Bolwig, N. (1959). A study of the behaviour of the Chacma baboon. *Behaviour*, **14**, 136–63.

Bookstein, F. L. (1978). *The Measurement of Biological Shape and Shape Change*. Berlin: Springer.

Botha, J. L., Bradshaw, D. & Gonin, R. (1985). Who dies of what? In *Proceedings of the N.A.M.D.A. Conference*, ed. A. B. Zwi & L. D. Saunders, pp. 41–54. Johannesburg: Hendlers.

Brace, C. L. (1967). Environment, tooth form and size in the Pleistocene. *Journal of Dental Research*, **46**, 809–16.

Brace, C. L. (1976). Krapina, 'classic' Neanderthals, and the evolution of the European face. *Journal of Human Evolution*, **8**, 527–50.

Brace, C. L. (1979). Tooth reduction in the Orient. *Asian Perspectives*, **19**, 203–19.

Brace, C. L. (1980). Australian tooth-size clines and the death of a stereotype. *Current Anthropology*, **21**, 141–53.

Brace, C. L. & Hinton, R. J. (1981). Oceanic tooth size variation as a reflection of biological and cultural mixing. *Current Anthropology*, **22**, 549–57.

Brace, C. L. & Mahler, P. E. (1971). Post-Pleistocene changes in the human dentition. *American Journal of Physical Anthropology*, **34**, 191–204.

Brace, C. L. & Montagu, M. F. A. (1965). *Man's Evolution*. New York: Macmillan.

Brace, C. L. & Nagai, M. (1982). Japanese tooth size, past and present. *American Journal of Physical Anthropology*, **59**, 399–411.

Brace, C. L., Rosenberg, K. R. & Hunt, K. D. (1987). Gradual change in human tooth size in the Late Pleistocene and Post-Pleistocene. *Evolution*, **41**, 705–20.

Brace, C. L. & Ryan, A. S. (1980). Sexual dimorphism and human tooth size. *Journal of Human Evolution*, **9**, 417–35.

Branson, R. S., Vaucher, Y. E., Harrison, G. G., Vargas, M. & Thies, C. (1982). Inter- and intra-observer reliability of skinfold thickness measurements in newborn infants. *Human Biology*, **54**, 137–43.

Brose, D. S. & Wolpoff, M. (1971). Early upper paleolithic man and late Middle Pleistocene teeth. *American Anthropologist*, **73**, 1156–94.

Brown, T. (1981). Comment on Brace & Hinton. *Current Anthropology*, **22**, 558.

Brown, T., Margetts, B. & Townsend, G. C. (1980). Comparison of MD crown diameters of the deciduous and permanent teeth in Australian Aborigines. *Australian Dental Journal*, **25**, 28–33.

Brown, T. & Townsend, G. C. (1980). Comment on Brace. *Current Anthropology*, **21**, 153.

Buettner-Janusch, I. (1966). *Origins of Man*. New York: John Wiley.

Bulmer, M. G. (1970). *The Biology of Twinning in Man*. Oxford: Clarendon.

Burdi, A. R., Garn, S. M. & Miller, R. L. (1970). Developmental advancement of the male dentition in the first trimester. *Journal of Dental Research*, **49**, 889.

Burian, R. M. (1983). Adaptation. In *Dimensions of Darwinism*, ed. M. Grene, pp. 287–314. Cambridge: Cambridge University Press.

Buschang, P. H., Tanguay, R. & Demirjian, A. (1987). Cephalometric reliability. *Angle Orthodontist*, **58**, 168–75.

Butler, P. M. (1937). Studies in the mammalian dentition – and of teeth of *Centetes ecaudatus* and its allies. *Proceedings of the Zoological Society, London, B*, **107**, 103–32.

Butler, P. M. (1939). Studies in the mammalian dentition – and of differentiation of the postcanine dentition. *Proceedings of the Zoological Society, London, B*, **109**, 1–36.

Calcagno, J. M. (1986). Dental reduction in post-Pleistocene Nubia. *American Journal of Physical Anthropology*, **70**, 349–63.

Cain, A. J. (1958). Logic and memory in Linnaeus' system of taxonomy. *Proceedings of the Linnaean Society of London*, **170**, 185–217.

Cameron, N. (1984). *The Measurement of Human Growth*. London: Croom Helm.

Campbell, N. A. (1978). Multivariate analysis in biological anthropology: some further considerations. *Journal of Human Evolution*, **7**, 197–203.

Campbell, N. A. (1984). Some aspects of allocation and discrimination. In *Multivariate Statistical Methods in Physical Anthropology*, ed. G. N. van Vark & W. W. Howells, pp. 177–92. Dordrecht: Reidl.

Campbell, N. A. & Atchley, W. R. (1981). The geometry of canonical variate analysis. *Systematic Zoology*, **30**, 268–80.

Canese, A. (1978). Datos actualizados sobre conocimientos epidemiologicos de Enfermedad de Chagas en el Paraguay. *Revista Paraguaya Microbiologica*, **13**, 93–6.

Canese, A. & Brice, E. (1978). Elevado indice de serologica positiva para la Enfermedad de Chagas en el Chaco Paraguaya. *Revista Paraguaya Microbiologica*, **13**, 3–6.

Cannon, W. B. (1935). Stresses and strains of homeostasis. *American Journal of Medical Science*, **189**, 1–14.

Carson, H. L. (1986). Patterns of inheritance. *American Zoologist*, **26**, 797–809.

Cavalli-Sforza, L. L. & Feldman, M. W. (1973). Cultural versus biological inheritance. *American Journal of Human Genetics*, **25**, 618–37.

Chamla, M. (1980). Etudes des variations métriques des couronnes dentaires des Nord-Africains de l'epipaleolithique à l'époque actuelle. *L'Anthropologie*, **84**, 254–71.

Chetverikov, S. S. (1926). On certain aspects of the evolutionary process from the standpoint of modern genetics. *Proceedings of the American Philosophical Society*, **105**, 167–95.

Cheverud, J. M. (1982a). Phenotypic, genetic and environmental morphological interaction in the cranium. *Evolution*, **36**, 499–516.

Cheverud, J. M. (1982b). Relationships among ontogenetic, static and evolutionary allometry. *American Journal of Physical Anthropology*, **59**, 139–49.

Cloninger, C. R., Rice, J. & Reich, T. (1979a). Multifactorial inheritance with cultural transmission and assortative mating. II. A general model. *American Journal of Human Genetics*, **31**, 179–98.

Cloninger, C. R., Rice, J. & Reich, T. (1979b). Multifactorial inheritance with cultural transmission and assortative mating. III. Family structure and the analysis of separation. *American Journal of Human Genetics*, **31**, 366–88.

Clutton-Brock, T. H. (1985). Size, sexual dimorphism and polygyny in primates. In

*Size and Scaling in Primate Biology*, ed. W. L. Jungers, pp. 51–60. New York: Plenum.

Clutton-Brock, T. H. & Harvey, P. H. (1976). Evolutionary rules and primate societies. In *Growing Points in Ethology*, ed. P. P. G. Bateson & R. A. Hinde, pp. 195–237. Cambridge: Cambridge University Press.

Clutton-Brock, T. H. & Harvey, P. H. (1979). Comparison and adaptation. *Proceedings of the Royal Society, London, B*, **205**, 547–65.

Cochard, L. R. (1985). Ontogenetic allometry of the skull and dentition of the Rhesus monkey *Macaca mulatta*. In *Size and Scaling in Primate Biology*, ed. W. L. Jungers, pp. 231–55. New York: Plenum.

Cock, A. G. (1966). Genetic aspects of metrical growth and form in animals. *Quarterly Review of Biology*, **41**, 131–90.

Cohen, J., Oliver, C. & Bernick, S. (1942). Dental studies in triplets. *Journal of Dental Research*, **21**, 233–9.

Cohen, M. M., Garn, S. M. & Geciauskas, M. A. (1970). Crown size profile pattern in trisomy. *Journal of Dental Research*, **49**, 460.

Cooley, W. W. & Lohnes, P. R. (1971). *Multivariate Data Analysis*. New York: Wiley.

Cooper, V. K. & Ludwig, T. G. (1965). Effect of fluoride and soil trace elements on the morphology of the permanent molars in man. *New Zealand Dental Journal*, **61**, 33–40.

Corruccini, R. S. (1972). Allometric correction in taximetrics. *Systematic Zoology*, **21**, 375–83.

Corruccini, R. S. (1973). Size and shape similarity coefficients based on metric characters. *American Journal of Physical Anthropology*, **38**, 743–54.

Corruccini, R. S. (1975). Multivariate analysis in biological anthropology: some considerations. *Journal of Human Evolution*, **4**, 1–19.

Corruccini, R. S. (1984). An epidemiologic transition in dental occlusion in world populations. *American Journal of Orthodontics*, **86**, 419–26.

Corruccini, R. S. (1987). Interpretation of metrical variables in multivariate analysis. In *Multivariate Statistical Methods in Physical Anthropology*, ed. G. N. Van Vark & W. W. Howells, pp. 13–19. Dordrecht: Reidl.

Corruccini, R. S. & Lee, G. T. R. (1984). Occlusal variation in Chinese immigrants to the United Kingdom and their offspring. *Archives of Oral Biology*, **29**, 779–82.

Corruccini, R. S. & McHenry, H. M. (1979). Morphological affinities of *Pan paniscus*. *Science*, **204**, 1341–3.

Cuvier, Baron. (1834). *The Animal Kingdom Arranged According to its Affinities*. London: Henderson.

Dahlberg, A. A. (1945). The changing dentition of man. *Journal of the American Dental Association*, **32**, 676–90.

Dahlberg, A. A. (1963). Dental evolution and culture. *Human Biology*, **35**, 237–49.

Dahlberg, A. A. (1986). Ontogeny and dental genetics in forensic problems. *Forensic Science International*, **30**, 163–76.

Dahlberg, G. (1940). *Statistical Methods for Medical and Biological Students*. London: Allen and Unwin.

Darwin, C. (1872). *The Origin of Species*. (6th edition). London: John Murray.

Darwin, C. (1899). *The Descent of Man and Selection in Relation to Sex*. (2nd edition). London: John Murray.

Darwin, C. (1905). *The Variation of Animals and Plants Under Domestication*. London: John Murray.

Davis, D. R. & Preble, J. S. (1986). Accuracy of hydrophilic irreversible hydrocolloid-silicone impression material. *Journal of Prosthetic Dentistry*, **55**, 304–8.

Day, B. & Keyser, R. (1987). *A graphical S.A.S. macro for statistical analysis of measurement method comparison and measurement repeatability*. Cary, N. C.: S.A.S.

de Terra, M. (1905). Beitrage zu einer Odontographie der Menschenrassen. Quoted by Miyabara, 1916.

de Vore, I. & Hall, K. R. L. (1965). Baboon ecology. In *Primate Behaviour*, ed. I. de Vore, pp. 20–52. New York: Holt, Rinehart and Winston.

Di Bennardo, R. & Bailit, H. L. (1978). Stress and dental asymmetry in a population of Japanese children. *American Journal of Physical Anthropology*, **48**, 89–94.

Dijkman, D. J. (1978). Patterns of Mesiodistal Toothsize Alteration. MA Thesis, University of Toronto.

Ditch, L. E. & Rose, J. C. (1972). A multivariate dental sexing technique. *American Journal of Physical Anthropology*, **37**, 61–4.

Doran, G. A. & Freedman, L. (1974). Metrical features of the dentition and arches of populations from Goroka and Lufa, Papua, New Guinea. *Human Biology*, **46**, 583–94.

Dorner, G. & Kawakami, M. (1978). *Hormones and Brain Development*. Amsterdam: Elsevier.

Doyle, W. J. & Johnston, O. (1977). On the meaning of increased fluctuating dental asymmetry: a cross-population study. *American Journal of Physical Anthropology*, **46**, 127–34.

Doyle, W. J., Kelley, C. & Siegel, M. I. (1977). The effects of audiogenic stress on the growth of long bones in the laboratory rat. *Growth*, **41**, 183–9.

Drennan, M. R. (1929). The dentition of a Bushman tribe. *Annals of the South African Museum*, **24**, 61–87.

Dullemeijer, P. (1980). Functional morphology and evolutionary biology. *Acta Biothermica*, **29**, 151–250.

Dunn, G. & Everitt, B. S. (1982). *An Introduction to Mathematical Taxonomy*. Cambridge: Cambridge University Press.

Dutta, P. C. (1983). An odontometric analysis of molar crown characters of Bronze Age Harappans. *Anthropologischer Anzeiger*, **41**, 67–72.

Ebeling, C. F., Ingervall, B., Hedegard, B. & Lewin, T. (1973). Secular changes in tooth size in Swedish men. *Acta Odontologica Scandinavica*, **31**, 140–7.

Efron, B. (1979). Bootstrap methods: another look at the jackknife. *Annals of Statistics*, **7**, 1–26.

Elston, R. C. & Stewart, J. (1971). A general model for the genetic analysis of pedigree data. *Human Heredity*, **21**, 523–42.

Evans, M. W. (1944). Congenital dental defects in infants subsequent to maternal rubella, during infancy. *Medical Journal of Australia*, **31**, 225–8.

Evans, M. W. (1947). Further observations on dental defects in infants subsequent to maternal rubella. *Medical Journal of Australia*, **1**, 780–5.

Falconer, D. S. (1967). *Quantitative Genetics*. Edinburgh: Oliver and Boyd.

Falk, D. & Corruccini, R. (1982). Efficacy of cranial versus dental measurements for separating human populations. *American Journal of Physical Anthropology*, **57**, 123–7.

Fanning, E. A. (1961). A longitudinal study of tooth formation and root resorption. *New Zealand Dental Journal*, **57**, 202–17.

Fedigan, L. M. (1982). *Primate Paradigms*. Montreal: Eden Press.

Filipson, R. & Goldson, L. (1963). Correlation between tooth width, width of head, length of head and stature. *Acta Odontologica Scandinavica*, **21**, 359–65.

Fisher, R. A. (1918). The correlation between relatives on the supposition of Mendelian inheritance. *Transactions of the Royal Society, Edinburgh*, **32**, 399–433.

Fisher, R. A. (1922). On the dominance ratio. *Proceedings of the Royal Society, Edinburgh*, **42**, 321–41.

Fisher, R. A. (1930). *The Genetical Theory of Natural Selection*. Oxford: Clarendon.

Fleagle, J. F. (1985). Size and adaptation in primates. In *Size and Scaling in Primate Biology*, ed. W. L. Jungers, pp. 1–19. New York: Plenum.

Ford, S. M. & Corruccini, R. S. (1985). Intraspecific, interspecific, metabolic and phylogenetic scaling in Platyrrhine primates. In *Size and Scaling in Primate Biology*, ed. W. L. Jungers, pp. 401–35. New York: Plenum.

Fowler, W. H. (1885). On the size of the teeth as a character of race. *Journal of the Royal Anthropological Institute*, **6**, 104–16.

Frayer, D. W. (1978). *Evolution of the dentition in the upper paleolithic and mesolithic Europe*. University of Kansas Publications in Anthropology, 10. Lawrence: Kansas University Press.

Freedman, L. (1962). Growth of muzzle length relative to calvaria length in Papio. *Growth*, **26**, 117–28.

Freedman, L. & Lofgren, M. (1981). Odontometrics of Western Australian Aborigines. *Archaeology in Oceania*, **16**, 87–93.

Furuhata, T. & Yamamoto, K. (1967). *Forensic Odontology*. Springfield (Mass.): C. C. Thomas.

Galton, F. (1876). A theory of heredity. *Journal of the Anthropological Institute of Great Britain*, **5**, 329–48.

Galton, F. (1883). Quoted by Tobias, 1955.

Garn, S. M. (1977). Genetics of dental development. In *Biology of Occlusal Development*, ed. J. A. McNamara, pp. 13–28. Ann Arbor: Center for Human Growth and Development.

Garn, S. M., Cohen, M. M. & Geciauskas, M. A. (1970). Increased crown-size asymmetry in Trisomy Y. *Journal of Dental Research*, **49**, 465.

Garn, S. M., Cole, P. E. & Smith, B. H. (1979). The effect of sample size on crown size asymmetry. *Journal of Dental Research*, **58**, 2012.

Garn, S. M., Lewis, A. B. & Bonne, B. (1961). Third molar polymorphism and the timing of tooth formation. *Nature*, **192**, 989.

Garn, S. M., Lewis, A. B. & Kerewsky, R. S. (1963a). Third molar polymorphism and its significance to dental genetics. *Journal of Dental Research*, **42**, 1344–63.

Garn, S. M., Lewis, A. B. & Kerewsky, R. S. (1963b). Third molar agenesis and size reduction of the remaining teeth. *Nature*, **200**, 488–9.

Garn, S. M., Lewis, A. B. & Kerewsky, R. S. (1963c). Phylogenetic and intraspecific variations in tooth sequence polymorphism. In *Dental Anthropology*, ed. D. R. Brothwell, pp. 53–73. New York: Pergamon.

Garn, S. M., Lewis, A. B. & Kerewsky, R. S. (1964a). Third molar agenesis and variation and size of the remaining teeth. *Nature*, **201**, 839.

Garn, S. M., Lewis, A. B. & Kerewsky, R. S. (1964b). Sex differences in tooth size. *Journal of Dental Research*, **43**, 306.

Garn, S. M., Lewis, A. B. & Kerewsky, R. S. (1965a). Size interrelationships of the mesial and distal teeth. *Journal of Dental Research*, **44**, 350–4.

Garn, S. M., Lewis, A. B. & Kerewsky, R. S. (1965b). Genetic, nutritional and maturational correlates of dental development. *Journal of Dental Research*, **44**, 228–42.

Garn, S. M., Lewis, A. B. & Kerewsky, R. S. (1965c). X-linked inheritance of tooth size. *Journal of Dental Research*, **44**, 439–41.

Garn, S. M., Lewis, A. B. & Kerewsky, R. S. (1966). The meaning of bilateral asymmetry in the permanent dentition. *Angle Orthodontist*, **36**, 55–62.

Garn, S. M., Lewis, A. B. & Kerewsky, R. S. (1967a). Shape similarities throughout the dentition. *Journal of Dental Research*, **46**, 1481.

Garn, S. M., Lewis, A. B. & Kerewsky, R. S. (1967b). Buccolingual size asymmetry and its developmental meaning. *Angle Orthodontist*, **37**, 186–93.

Garn, S. M., Lewis, A. B. & Kerewsky, R. S. (1968a). The magnitude and implications of the relationship between tooth size and body size. *Archives of Oral Biology*, **13**, 129–31.

Garn, S. M., Lewis, A. B. & Kerewsky, R. S. (1968b). Relationships between buccolingual and mesiodistal tooth diameters. *Journal of Dental Research*, **47**, 495.

Garn, S. M., Lewis, A. B., Swindler, D. R. & Kerewsky, R. S. (1967c). Genetic control of sexual dimorphism in tooth size. *Journal of Dental Research*, **46**, 963–72.

Garn, S. M., Lewis, A. B. & Walenga, A. (1968c). Evidence for a secular trend in tooth size over two generations. *Journal of Dental Research*, **47**, 503.

Garn, S. M., Osborne, R. H. & McCabe, K. D. (1979). The effect of prenatal factors on crown dimensions. *American Journal of Physical Anthropology*, **51**, 665–78.

Garn, S. M. & Rohmann, G. C. (1962). X-linked inheritance of developmental timing in man. *Nature*, **196**, 695–6.

Garn, S. M., Smith, B. H. & Moyers, R. E. (1981). Structured (patterned) dimensional and developmental dental asymmetry. *Proceedings of the Finnish Dental Society*, **77**, 33–6.

Ghose, L. J. & Bahdady, V. S. (1979). Analysis of the Iraqi dentition: mesiodistal diameters of permanent teeth. *Journal of Dental Research*, **58**, 1047–54.

Gingerich, P. D. (1977). Correlation of tooth size and body size in living hominid primates with a note on relative brain size in *Aegyptopithecus* and *Proconsul*. *American Journal of Physical Anthropology*, **47**, 395–8.

Gingerich, P. D. & Schoeninger, M. J. (1979). Patterns of tooth size variability in the dentition of primates. *American Journal of Physical Anthropology*, **51**, 457–66.

Gingerich, P. D. & Smith, M. J. (1985). Allometric scaling in the dentition of primates and insectivores. In *Size and Scaling in Primate Biology*, ed. W. L. Jungers, pp. 257–72. New York: Plenum.

Gingerich, P. D., Smith, M. J. & Rosenberg, K. (1982). Allometric scaling in the dentition of primates and prediction of body weight from tooth size in fossils. *American Journal of Physical Anthropology*, **58**, 81–100.

Goldberg, S. (1929). Biometrics of identical twins from a dental viewpoint. *Journal of Dental Research*, **9**, 363–409.

Goldberg, S. (1939). The dental arcades of identical twins. *Dental Cosmos*, **72**, 869–81.

Goldstein, S., Post, D. & Melnick, D. (1978). An analysis of cercopithecoid odontometrics. I. The scaling of the maxillary dentition. *American Journal of Physical Anthropology*, **49**, 517–32.

Goose, D. H. (1963). Dental measurement: an assessment of its value in anthropological studies. In *Dental Anthropology*, ed. D. R. Brothwell, pp. 125–48. New York: Pergamon.

Goose, D. H. (1967). Preliminary study of tooth size in families. *Journal of Dental Research*, **46**, 959–62.

Goose, D. H. (1971). The inheritance of tooth size in British families. In *Dental Morphology and Evolution*, ed. A. A. Dahlberg, pp. 144–9. Chicago: University of Chicago Press.

Goose, D. H. & Appleton, J. (1982). *Human Dentofacial Growth*. New York: Pergamon.

Gosset, W. S. (1908). The Student *t*-test. In *Statistics at Square One*, ed. T. D. V. Swinscow, pp. 14–31. London: British Medical Association.

Gould, S. J. (1966). Allometry and size in ontogeny and phylogeny. *Biological Reviews*, **41**, 587–640.

Gould, S. J. (1975). Allometry in primates, with emphasis on scaling and the evolution of the brain. In *Approaches to Primate Paleobiology*, ed. F. Szalay, pp. 244–92. Basel: Karger.

Gould, S. J. (1977). *Ontogeny and Phylogeny*. Cambridge: Harvard University Press.

Gould, S. J. (1978). Generality and uniqueness in the history of life: and exploration with random models. *Bioscience*, **28**, 277–81.

Gould, S. J. (1981). *The Mismeasure of Man*. New York: W. W. Norton.

Gould, S. J. & Garwood, R. A. (1969). Levels of integration in the mammalian dentition. *Evolution*, **23**, 276–300.

Gould, S. J. & Lewontin, R. (1979). The spandrels of San Marco and the Panglossan paradigm: a critique of the adaptationist programme. *Proceedings of the Royal Society, London, B*, **205**, 581–98.

Gould, S. J. & Vrba, E. S. (1982). Exaptation – a missing term in the science of form. *Paleobiology*, **8**, 4–15.

Grant, D. A., Stern, I. B. & Everett, F. G. (1979). *Periodontics*. Mosby: St Louis.

Grassi, J. & Salinas, V. (1979). Measles in Paraguay. *Bulletin of the Pan American Health Organization*, **13**, 139–46.

Greene, D. L. (1984). Fluctuating dental asymmetry and measurement error. *American Journal of Physical Anthropology*, **65**, 283–9.

Grewal, M. S. (1962). Development of an inherited tooth defect in the mouse. *Journal of Embryology and Experimental Morphology*, **10**, 202–11.

Grine, F. E. (1981). Relative sizes of the maxillary deciduous canine and central incisor teeth in the Kalahari San and South African Negro. *Annals of the South African Museum*, **7**, 229–45.

Groeneveld, H. T. & Kieser, J. A. (1987). An evaluation of the M- statistic in human odontomorphometric distance analyses. *International Journal of Anthropology*, **2**, 29–36.

Groeneveld, H. T. & Kieser, J. A. (1988). Confidence limits for Penrose's shape coefficient in *Cercopithecus aethiops*: a simulation approach. *International Journal of Primatology*, **9**, 49–58.

Gruneberg, H. (1951). The genetics of a tooth defect in the mouse. *Proceedings of the Royal Society, London, B*, **138**, 437–51.

Gustafson, G. (1966). *Forensic Odontology*. New York: Elsevier.

Hampé, A. (1960). La Competition entre les Elements Osseux du Zengopode de Poulet. *Journal of Embryology and Experimental Morphology*, **8**, 241–5.

Hanihara, K. (1970). Upper lateral incisor variability and the size of remaining teeth. *Journal of the Anthropological Society of Nippon*, **78**, 316–23.

Hanihara, K. (1976). Statistical and comparative studies of the Australian Aboriginal dentition. *Bulletin of the University Museum, Tokyo*, **11**, 1–57.

Hanihara, K. (1977). Distances between Australian Aborigines and certain other populations based on dental measurements. *Journal of Human Evolution*, **6**, 403–18.

Hanihara, K., Masuda, T. & Tanaka, T. (1965). Evolutionary significance of reduced and supernumerary teeth in the dentition. *Journal of the Anthropological Society of Nippon*, **73**, 72–80.

Hanihara, K. & Ueda, H. (1979). Crown diameters in Japanese–American F1 hybrids. *Ossa*, **6**, 105–14.

Harris, E. F. & Bailit, H. L. (1988). A principal components analysis of human odontometrics. *American Journal of Physical Anthropology*, **75**, 87–99.

Harris, E. F. & Nweeia, M. T. (1980a). Dental asymmetry as a measure of environmental stress in the Ticuna Indians of Columbia. *American Journal of Physical Anthropology*, **53**, 133–42.

Harris, E. F. & Nweeia, M. T. (1980b). Tooth size of Ticuna Indians, Colombia, with phenetic comparisons to other Amerindians. *American Journal of Physical Anthropology*, **53**, 81–91.

Harrison, G. A., Weiner, J. S., Tanner, J. M. & Barnicot, N. A. (1964). *Human Biology*. London: Oxford University Press.

Harvey, P. H., Kavanagh, M. & Clutton-Brock, T. H. (1978). Sexual dimorphism in primate teeth. *Journal of Zoology*, **186**, 475–85.

Harzer, W. (1987). A hypothetical model of genetic control of tooth crown growth in man. *Archives of Oral Biology*, **32**, 159–62.

Hasund, A. P. (1966). *Okklusion og facialkranium i middelalderbefolkigen i Oslo of Heidal*. Oslo: Anatomik Institutt, Audeling for Anthropologi, University of Oslo.

Haugen, L. K. (1977). The human upper and middle face: a morphological investigation. I. Introduction, materials and methods. *Acta Morphologica Neerdandia Scandinavica*, **15**, 241–68.

Hellman, M. (1936). Our third molar teeth, their eruption, presence and absence. *Dental Cosmos*, **78**, 750–62.

Henderson, A. M. & Corruccini, R. S. (1976). Relationship between tooth size and body size in American Blacks. *Journal of Dental Research*, **55**, 94–6.

Henderson, A. M. & Greene, D. L. (1975). Dental field theory: an application to primate evolution. *Journal of Dental Research*, **54**, 344–50.

Herring, S. W. (1972). The role of the canine morphology in the evolutionary divergence of pigs and peccaries. *Journal of Mammalogy*, **53**, 500–12.

Hersch, A. H. (1955). Allometry and anthropometry. *Annals of the New York Academy of Science*, **63**, 484–90.

Hinkle, L. E. (1973). The concept of stress in the biological and social sciences. *Science, Medicine and Man*, **1**, 31–48.

Hinton, R. J., Smith, M. O. & Smith, F. H. (1980). Tooth size changes in prehistoric Tennessee Indians. *Human Biology*, **52**, 229–45.

Hollinger, J. O., Lorton, L., Krantz, W. A. & Connelly, M. (1984). A clinical and laboratory comparison of irreversible hydrocolloid impression techniques. *Journal of Prosthetic Dentistry*, **51**, 304–09.

Holloway, P. J., Shaw, J. H. & Sweeney, E. A. (1961). Effects of various sucrose: casein ratios in purified diets on the teeth and supporting structures of rats. *Archives of Oral Biology*, **3**, 185–200.

Holloway, R. L. (1966). Structural reduction through the probable mutation effect. *American Journal of Physical Anthropology*, **27**, 7–12.

Horowitz, S. L., Osborne, R. H. & de George, F. V. (1958). Hereditary factors in tooth dimensions, a study of the anterior teeth in twins. *Angle Orthodontist*, **28**, 87–93.

Howells, W. W. (1984). Introduction. In *Multivariate Statistical Methods in Physical Anthropology*, ed. G. N. van Vark & W. W. Howells, pp. 1–11. Dordrecht: Reidl.

Hrdlička. (1911). Human dentition and teeth from the evolutionary and racial standpoint. *Dominion Dental Journal*, **23**, 403–21.

Hrdlička. (1952). *Practical Anthropometry*. Philadelphia: Wistar.

Hughes, A. J. & Lambert, D. M. (1984). Functionalism, structuralism and ways of seeing. *Journal of Theoretical Biology*, **111**, 787–800.

Hunter, W. S. & Priest, W. R. (1960). Errors and discrepancies in measurement of tooth size. *Journal of Dental Research*, **39**, 405–14.

Huxley, J. S. (1932). *The Problems of Relative Growth*. London: Methuen.

Huxley, J. S. & de Beer, G. R. (1934). *Elements of Experimental Embryology*. Cambridge: Cambridge University Press.

Hylander, W. (1975). Incisor size and diet in anthropoids with special reference to *Cercopithecidae*. *Science*, **189**, 1095–8.

Ingervall, B., Lewin, T. Hedegard, B. (1972). Secular changes in the morphology of the skull in Swedish men. *Acta Odontologica Scandinavica*, **30**, 539–41.

Irwig, L. M. & Ingle, R. (1984). Childhood mortality rates, infant feeding and the use of health services in rural Transkei. *South African Medical Journal*, **66**, 607–13.

Jacobson, A. (1982). *The Dentition of the South African Negro*. Anniston (Alabama): Higginsbotham.

Johannsen, W. (1909). *Elemente der Exakten Erblichkeitslehre*. Jena: Gustav Fischer.

Johnston, F. E. (1980). Research design and sample selection. In *Human Physical Growth and Maturation*, ed. F. E. Johnston, A. F. Roche & C. Suzanne, pp. 5–19. New York: Plenum.

Jordan, H. E. (1914). Hereditary left handedness with a note on twinning. *Journal of Genetics*, **4**, 67–81.

Jordan, S. D. & Evermann, B. W. (1896). The fishes of North and Middle America. *Bulletin of the United States National Museum*, **47**, 1–393.

Joysey, K. A. (1959). A study of variation and relative growth in the Blastoid Orbitremites. *Philosophical Transactions of the Royal Society, London, B*, **242**, 99–125.

Jungers, W. L. (1978). On canines in early hominids. *Current Anthropology*, **19**, 155–6.

Katz, M. J. (1982). Ontogenetic mechanisms: the middle ground of evolution. In *Evolution and Development*, ed. J. T. Bonner, pp. 207–13. Berlin: Springer.

Kay, R. F. (1975a). The functional adaptations of primate molar teeth. *American Journal of Physical Anthropology*, **43**, 195–216.

Kay, R. F. (1975b). Allometry and early hominids. *Science*, **189**, 63.

Kay, R. F. (1978). Molar structure and diet in extant *Cercopithecidae*. In

*Development, Function and Evolution of Teeth*, ed. P. Butler & K. Joysey, pp. 309–39. New York: Academic Press.

Kay, R. F. & Simmons, E. L. (1980). The ecology of Oligocene African Anthropoideae. *International Journal of Primatology*, 1, 21–37.

Kieser, J. A. (1985). An odontometric analysis of the early Griqua dentition. *Anthropologischer Anzeiger*, 43, 51–8.

Kieser, J. A. (1986). Odontogenic polarity and Butler's field theory. *Medical Hypotheses*, 20, 103–7.

Kieser, J. A. & Groeneveld, H. T. (1986a). Fluctuating odontometric asymmetry in a South African caucasoid population. *Journal of the Dental Association of South Africa*, 41, 185–9.

Kieser, J. A. & Groeneveld, H. T. (1986b). The assessment of fluctuating odontometric asymmetry from incomplete hominid data. *Anthropologischer Anzeiger*, 44, 175–82.

Kieser, J. A. & Groeneveld, H. T. (1987a). A metrical analysis of the *Cercopithecus aethiops* dentition. *International Journal of Primatology*, 8, 261–80.

Kieser, J. A. & Groeneveld, H. T. (1987b). Patterns of metrical variability in the dentition of *Papio ursinus*. *American Journal of Primatology*, 14, 141–51.

Kieser, J. A. & Groeneveld, H. T. (1987c). Tooth size and arcadal length correlates in man. *International Journal of Anthropology*, 2, 37–46.

Kieser, J. A. & Groeneveld, H. T. (1987d). Static intraspecific allometry of jaws and teeth in *Cercopithecus aethiops*. *Journal of Zoology*, 212, 499–510.

Kieser, J. A. & Groeneveld, H. T. (1988a). Static intraspecific maxillo-facial allometry in the chacma baboon. *Folia Primatologica*, 48, 151–63.

Kieser, J. A. & Groeneveld, H. T. (1988b). Patterns of variability in the South African Negro dentition. *Journal of the Dental Association of South Africa*, 43, 105–10.

Kieser, J. A. & Groeneveld, H. T. (1988c). Fluctuating dental asymmetry in an urban South African black population. *Journal of Dental Research*, 67, 1200–5.

Kieser, J. A. & Groeneveld, H. T. (1988d). Meristic variation and odontogenetic stability in man. *South African Journal of Science*, 84, 209–12.

Kieser, J. A. & Groeneveld, H. T. (1988e). Aspects of the human cranial base: a multivariate approach. In *Tobias Festschrift*, ed. G. H. Sperber, (in press). New York: Alan Liss.

Kieser, J. A. & Groeneveld, H. T. (1988f). Allocation and discrimination based on human odontometric data. *American Journal of Physical Anthropology*, (in press).

Kieser, J. A. & Groeneveld, H. T. (1988g). Allometric relations of teeth and jaws in man. *American Journal of Physical Anthropology*, 77, 57–68.

Kieser, J. A., Groeneveld, H. T. & Cameron, N. (1987). Evidence for a secular trend in the Negro dentition. *Annals of Human Biology*, 14, 517–32.

Kieser, J. A., Groeneveld, H. T. & Preston, C. B. (1985a). Age related tooth size variation in the Lengua Indians of Paraguay. *Zeitschrift für Morphologie und Anthropologie*, 75, 341–8.

Kieser, J. A., Groeneveld, H. T. & Preston, C. B. (1985b). Craniofacial correlates of attrition. *Angle Orthodontist*, 55, 329–35.

Kieser, J. A., Groeneveld, H. T. & Preston, C. B. (1985c). Patterns of dental wear in the Lengua Indians of Paraguay. *American Journal of Physical Anthropology*, 66, 21–9.

Kieser, J. A., Groeneveld, H. T. & Preston, C. B. (1985d). An odontometric analysis of the Lengua Indians dentition. *Human Biology*, **57**, 611–20.

Kieser, J. A., Groeneveld, H. T. & Preston, C. B. (1985e). A metrical analysis of the South African caucasoid dentition. *Journal of the Dental Association of South Africa*, **40**, 121–5.

Kieser, J. A., Groeneveld, H. T. & Preston, C. B. (1985f). Compensatory tooth size interactions in a preliterate population. *Journal of Human Evolution*, **14**, 739–45.

Kieser, J. A., Groeneveld, H. T. & Preston, C. B. (1986a). Fluctuating odontometric asymmetry in the Lengua Indians of Paraguay. *Annals of Human Biology*, **13**, 489–98.

Kieser, J. A., Groeneveld, H. T. & Preston, C. B. (1986b). Fluctuating dental asymmetry as a measure of odontogenetic canalization. *American Journal of Physical Anthropology*, **71**, 437–44.

Kieser, J. A., Groeneveld, H. T. & Preston, C. B. (1986c). On the non-existence of compensatory tooth size interaction in a contemporary human population. *Journal of Dental Research*, **65**, 1105–07.

Kieser, J. A., Preston, C. B. & Evans, W. G. (1983). Skeletal age at death: an evaluation of the Miles method of ageing. *Journal of Archaeological Science*, **10**, 9–12.

Kimura, M. & Ohta, T. (1973). Mutation and evolution at the molecular level. *Genetics*, **73**, 19–35.

Kimura, M. & Ohta, T. (1974). Some principles governing molecular evolution. *Proceedings of the National Academy of Science of the United States of America*, **71**, 2848–52.

Kirveskari. P., Hansson, H., Hedegard, B. & Karlson, N. (1977). Crown size and hypodontia in the permanent teeth of modern Skolt Lapps. *American Journal of Physical Anthropology*, **48**, 107–12.

Klecka, W. R. (1980). *Discriminant Analysis*. Beverley Hills:

Kolakowski, D. & Bailit, H.L. (1981a). A differential environmental effect on human anterior tooth size. *American Journal of Physical Anthropology*, **54**, 377–81.

Kolakowski, D. & Bailit, H. L. (1981b). Segregation analyses of tooth size in a Melanesian population. *Annals of Human Biology*, **8**, 255–62.

Korenhof, C. A. W. (1960). *Morphological Aspects of the Human Upper Molar*. Utrecht: Neederlandia.

Koritzer, R. T. (1980). Comment on Brace, 1980. *Current Anthropology*, **21**, 156.

Korkhaus, G. (1930). Anthropologic and odontologic studies in twins. *International Journal of Orthodontics*, **16**, 640–7.

Koyoumdjisky-Kaye, E., Steigman, S. & Gudelevitch, B. (1978). A comparative study of tooth and arch dimensions and sex dimorphism in Israeli children of Cochin and North African descent. *Zeitschrift für Morphologie und Anthropologie*, **69**, 32–42.

Koyoumdjisky-Kaye, E., Zilberman, Y. & Hazan, O. (1977). A comparative study of tooth and dental arch dimensions and sexual dimorphism in Circassian and Druze children. *Zeitschrift für Morphologie und Anthropologie*, **68**, 298–306.

Koyoumdjisky-Kaye, E., Zilberman, Y. & Zeevi, Z. (1976). A comparative study of tooth and arch dimensions in Jewish children of differing ethnic descent. *American Journal of Physical Anthropology*, **44**, 347–444.

Krogstad, O. (1972). Maxillary prognathism in the Greenland Eskimo. *Zeitschrift für Morphologie und Anthropologie*, **64**, 279–307.

Kurtén, B. (1957). A case of Darwinian selection in bears. *Evolution,* **11**, 412–16.

Kurtén, B. (1967). Some quantitative approaches to dental microevolution. *Journal of Dental Research,* **46**, 817–28.

Lande, R. (1985). Genetic and evolutionary aspects of allometry. In *Size and Scaling in Primate Biology,* ed. W. L. Jungers, pp. 21–32. New York: Plenum.

Lande, R. (1986). The dynamics of peak shifts and the pattern of evolution. *Paleobiology,* **12**, 343–54.

Lande, R. & Arnold, S. J. (1983). The measurement of selection on correlated characters. *Evolution,* **37**, 1210–26.

Larson, S. (1981). Organ weight scaling in four species of macaque. *American Journal of Physical Anthropology,* **54**, 243.

Lauer, C. (1975). The relationship between tooth size and body size in a population of rhesus monkeys, *Macaca mulatta. American Journal of Physical Anthropology,* **43**, 333–40.

Lavelle, C. L. B. (1972). Maxillary and mandibular tooth size in different racial groups and in different occlusal categories. *American Journal of Orthodontics,* **61**, 29–37.

Lavelle, C. L. B. (1977). Relationship between tooth and long bone size. *American Journal of Physical Anthropology,* **46**, 423–6.

Leamy, L. & Bradley, D. (1982). Static and growth allometry of morphometric traits in random bred mice. *Evolution,* **36**, 1200–12.

Leary, R. F., Allendorf, F. W. & Knudsen, K. L. (1985). Inheritance of meristic variation and the evolution of developmental stability. *Evolution,* **39**, 308–14.

Le Blanc, S. A. & Black, B. (1974). A long term trend in tooth size in the eastern Mediterranean. *American Journal of Physical Anthropology,* **41**, 417–22.

Le Bot, P., Gueguen, A. & Salmon, D. (1980). Congenital defects of the upper lateral incisors and the morphology of the other teeth in man. *American Journal of Physical Anthropology,* **53**, 479–86.

Le Bot, P. & Salmon, D. (1977). Congenital defects of the upper lateral incisor: condition and measurements of the upper teeth. *American Journal of Physical Anthropology,* **46**, 231–44.

Lee, G. T. R. & Goose, D. H. (1972). The dentition of Chinese living in Liverpool. *Human Biology,* **44**, 563–72.

Leutenegger, W. (1972). Review of: Metric trends in hominid evolution. *Social Biology,* **19**, 76–7.

Leutenegger, W. & Cheverud, J. M. (1985). Sexual dimorphism in primates: the effects of size. In *Size and Scaling in Primate Biology,* ed. W. L. Jungers, pp. 33–50. New York: Plenum.

Leutenegger, W. & Kelly, J. T. (1977). Relationship of sexual dimorphism in canine size and body size to social, behavioural and ecological correlates in anthropoid primates. *Primates,* **18**, 117–36.

Lewis, D. W. & Grainger, R. M. (1967). Sex-linked inheritance of tooth size. *Archives of Oral Biology,* **12**, 539–44.

Lewontin, R. C. (1974). *The Genetic Basis of Evolutionary Change.* New York: Columbia University Press.

Lewontin, R. C. (1986). How important is genetics for an understanding of evolution? *American Zoologist,* **26**, 811–20.

Liu, K. L. (1977). Dental condition of two tribes of Taiwan aborigines. *Journal of Dental Research,* **56**, 117–27.

Lombardi, V. (1975). A factor analysis of morphogenetic fields in the human dentition. *American Journal of Physical Anthropology*, **42**, 99–104.

Ludwig, W. (1932). *Das Rechts-Links Problem im Tierreich und beim Menschen.* Berlin: Springer.

Lucas, P. W. (1981). An analysis of the canine tooth size of old world higher primates. *Journal of Zoology*, **195**, 437–48.

Lucas, P. W. (1982). An analysis of the canine tooth size of old world higher primates in relation to mandibular length and body weight. *Archives of Oral Biology*, **27**, 493–6.

Lukacs, J. R. (1977). Anthropological aspects of dental variation in North India. PhD Thesis, Cornell University.

Lukacs, J. R. (1981). Crown dimensions of deciduous teeth from prehistoric India. *American Journal of Physical Anthropology*, **55**, 261–6.

Lukacs, J. R. (1985). Tooth size variation in prehistoric India. *American Anthropologist*, **87**, 811–25.

Lukacs, J. R., Joshi, M. R. & Makhija, P. G. (1983). Crown dimensions of deciduous teeth in prehistoric and living populations of West India. *American Journal of Physical Anthropology*, **61**, 383–7.

Lundström, A. (1948). *Tooth Size and Occlusion in Twins.* New York: Karger.

Lundström, A. (1954). Intermaxillary tooth width ratio, tooth alignment and occlusion. *Acta Odontologica Scandinavica*, **12**, 265–72.

Lundström, A. (1955). The significance of genetic and nongenetic factors in the profile and facial skeleton. *American Journal of Orthodontics*, **41**, 910–16.

Lundström, A. (1963). Tooth morphology as a basis for distinguishing monozygotic and dizygotic twins. *American Journal of Human Genetics*, **15**, 34–43.

Lundström, A. (1967). Genetic aspects of variation in tooth width based on symmetry and twin studies. *Hereditas*, **57**, 403–9.

Lunt, D. A. (1969). An odontometric study of mediaeval Danes. *Acta Odontologica Scandinavica*, **27**, 55–113.

Macchiarelli, R. & Bondioli, L. (1986). Post-Pleistocene reduction in human dental structure. *Human Evolution*, **1**, 405–18.

Margulis, L. & Sagan, D. (1984). Evolutionary origins of sex. In *Oxford Survey in Evolutionary Biology*, ed. R. Dawkins & M. Ridley, vol. 1, pp. 16–47. Oxford: Oxford University Press.

Mason, L. G., Ehrlich, P. R. & Emmel, T. C. (1967). The population biology of the butterfly, *Euphydras editha. V.* character clusters and asymmetry. *Evolution*, **21**, 85–91.

Mather, K. (1953). Genetic control of stability in development. *Heredity*, **7**, 297–336.

Mather, K. & Jinks, J. L. (1963). Correlations between relatives arising from sex linked genes. *Nature*, **198**, 314–15.

Mayhall, J. T. (1979). The dental morphology of the Inuit of the Canadian Central Arctic. *Ossa*, **6**, 199–218.

Mayr, E. (1966). *Animal Species and Evolution.* Cambridge: Belknap.

Mayr, E. (1976). *Evolution and the Diversity of Life.* Cambridge: Harvard University Press.

Mayr, E. (1982). *The Growth of Biological Thought.* Cambridge: Belknap.

McHenry, H. M. & Corruccini, R. S. (1981). *Pan paniscus* and human evolution. *American Journal of Physical Anthropology*, **54**, 355–67.

McKee, J. (1984). A genetic model of dental reduction through the probable mutation effect. *American Journal of Physical Anthropology*, **65**, 231–41.

McNamara, K. J. (1982). Heterochrony and phylogenetic trends. *Paleobiology*, **8**, 130–42.

Menezes, D. M., Foster, T. D. & Lavelle, C. L. B. (1974). Genetic influences on dentition and the dental arch dimensions. *American Journal of Physical Anthropology*, **40**, 213–20.

Mijsberg, W. A. (1931). On sexual differences in the teeth of Japanese. *Koninklijke Akademie voor Wetenschap*, **34**, 1111–15.

Miyabara, T. (1916). An anthropological study of the masticatory system in Japanese. I. The teeth. *Dental Cosmos*, **58**, 739–49.

Mizoguchi, Y. (1977). Genetic variability of permanent tooth crowns as ascertained from twin data. *Journal of the Anthropological Society of Nippon*, **85**, 301–9.

Mizoguchi, Y. (1980). Factor analysis of environmental variation in the permanent dentition. *Bulletin of the National Science Museum, Tokyo*, **6**, 29–46.

Mizoguchi, Y. (1981a). Comment on Brace, 1981. *Current Anthropology*, **22**, 560.

Mizoguchi, Y. (1981b). Variation units in the human permanent dentition. *Bulletin of the National Science Museum, Tokyo*, **7**, 29–39.

Mizoguchi, Y. (1983a). Influences of the earlier developing teeth upon later developing teeth. *Bulletin of the National Science Museum, Tokyo*, **9**, 33–45.

Mizoguchi, Y. (1983b). Confirmatory factor analysis by maximal likelihood method with respect to the influence of the developmental periods on the size of the teeth. *Report of the Ministry of Education, Science and Culture, Japan*, 1–32.

Mizoguchi, Y. (1986). Correlated asymmetries detected in tooth crown diameters of human permanent teeth. *Bulletin of the National Science Museum Tokyo*, **12**, 25–45.

Mizoguchi, Y. (1987). Mirror imagery and genetic variability of lateral asymmetries in the mesiodistal crown diameters of permanent teeth. *Bulletin of the National Science Museum, Tokyo*, **13**, 10–20.

Mizoguchi, Y. (1988). Statistical analysis of geographical variation in dental size. *Report of the Ministry of Education, Science and Culture, Japan*, pp. 1–124.

Mjor, I. A. & Hougen, E. (1974). Pulp structure in bilateral and opposing pairs of teeth. *Scandinavian Journal of Dental Research*, **82**, 128–34.

Molnar, S. (1971). Human tooth wear, tooth function and cultural variability. *American Journal of Physical Anthropology*, **34**, 175–90.

Moorrees, C. F. A. (1957). *The Aleut Dentition*. Cambridge: Harvard University Press.

Moorrees, C. F. A. (1959). *The Dentition of the Growing Child*. Cambridge: Harvard University Press.

Moorrees, C. F. A., Fanning, E. A. & Hunt, E. E. (1963). Age variation of formation stages in ten permanent teeth. *Journal of Dental Research*, **42**, 1490–502.

Moorrees, C. F. A. & Reed, R. B. (1954). Correlations among crown diameters of human teeth. *Archives of Oral Biology*, **9**, 685–97.

Moorrees, C. F. A., Thomsen, S. O., Jensen, E. & Yen, P. K. J. (1957). Mesiodistal crown diameters of deciduous and permanent teeth. *Journal of Dental Research*, **36**, 39–47.

Morton, N. E. (1974). Analysis of family resemblance. I. Introduction. *American Journal of Human Genetics*, **26**, 318–30.

Mosiman, J. E. (1965). Variation on relative growth in the plastral scutes of the turtle. *Miscellaneous Publications of the Museum of Zoology, University of Michigan*, **97**, 1–43.

Mosiman, J. E. & James, F. C. (1979). New statistical methods for allometry with application to Florida redwinged blackbirds. *Evolution*, **33**, 444–59.

Moss, M. L. (1973). A functional cranial analysis of primate craniofacial growth. In *Craniofacial Biology of Primates*. Proceedings of the Fourth International Congress of Primatology, **3**, 191–208.

Moss, M. L. (1978). Analysis of developmental processes possibly related to human dental sexual dimorphism. In *Development, Function and Evolution of Teeth*, ed. P. M. Butler & K. A. Joysey, pp. 135–48. New York: Academic Press.

Moss, M. L. & Chase, P. S. (1966). Morphology of Liberian Negro deciduous teeth. *American Journal of Physical Anthropology*, **24**, 215–30.

Moss, M. L., Chase, P. S. & Howes, R. I. (1967). Comparative odontometry of the permanent postcanine dentition of American Whites and Negroes. *American Journal of Physical Anthropology*, **27**, 125–42.

Moss, M. L. & Salentijn, L. (1969). The primary role of functional matrices in facial growth. *American Journal of Orthodontics*, **55**, 566–77.

Moss, M. L. & Salentijn, L. (1970). The logarithmic growth of the human mandible. *Acta Anatomica*, **77**, 341–60.

Moss, M. L. & Salentijn, L. (1971). The unitary logarithmic curve description of human mandibular growth. *Acta Anatomica*, **78**, 532–42.

Moss, M. L. & Young, R. (1960). A functional approach to craniology. *American Journal of Physical Anthropology*, **18**, 281–92.

Muhlreiter, E. (1892). Quoted by Wolpoff, 1969.

Nelson, C. T. (1938). The teeth of the Indians of Pecos Pueblo. *American Journal of Physical Anthropology*, **23**, 261–93.

Neville, A. C. (1976). *Animal Asymmetry*. London: Edward Arnold.

Niswander, J. D. & Chung, C. S. (1965). The effects of inbreeding on tooth size in Japanese children. *American Journal of Human Genetics*, **17**, 390–8.

Nolla, C. M. (1960). The development of the permanent teeth. *Journal of Dentistry for Children*, **27**, 254–66.

Odell, G., Oster, G., Burnside, B. & Alberch, P. (1980). The mechanical basis of morphogenesis. *Developmental Biology*, **85**, 1–17.

Oliver, G. (1960). *Pratique Anthropologique*. Paris: Feres.

Olson, E. & Miller, R. (1958). *Morphological Integration*. Chicago: Chicago University Press.

Onions, C. T. (1973). *The Shorter Oxford English Dictionary*. Oxford: Clarendon Press.

O'Rourke, D. H. & Crawford, M. H. (1980). Odontometric microdifferentiation of transplanted Mexican Indian populations. *American Journal of Physical Anthropology*, **52**, 421–34.

Osborn, J. W. (1970). A new approach to zahnreihen. *Nature*, **225**, 343–6.

Osborn, J. W. (1971). The ontogeny of tooth succession in *Lacerta vivipara*. *Proceedings of the Royal Society, London, B*, **179**, 261–89.

Osborn, J. W. (1973). The evolution of dentitions. *American Scientist*, **61**, 548–59.

Osborn, J. W. (1975). Tooth replacement: efficiency, patterns and evolution. *Evolution*, **29**, 180–6.

Osborn, J. W. (1978). Morphogenetic gradients: fields versus clones. In *Development, Function and Evolution of Teeth*, ed. P. M. Butler & K. A. Joysey, pp. 171–201. London: Academic Press.

Osborn, J. W. & Ten Cate, A. R. (1983). *Advanced Dental Histology*. Bristol: Wright.

Osler, W. (1910). *The Lumleian lectures on angina pectoris. Lancet*, i, 696–700.

Overall, J. E. & Klett, C. J. (1972). *Applied Multivariate Statistics*. New York: McGraw-Hill.

Paynter, K. J. & Grainger, R. M. (1956). The relation of nutrition to the morphology and size of rat molar teeth. *Journal of the Canadian Dental Association*, **22**, 519–31.

Pengilly, D. (1984). Developmental versus functional explanations for patterns of variability and correlation in the dentitions of foxes. *Journal of Mammalogy*, **65**, 34–43.

Penrose, L. S. (1954). Distance, size and shape. *Annals of Eugenics*, **18**, 337–43.

Perzigian, A. J. (1975). Natural selection on the dentition of an Arikara population. *American Journal of Physical Anthropology*, **42**, 63–70.

Perzigian, A. J. (1976). The dentition of the Indian Knoll skeletal population: odontometrics and cusp number. *American Journal of Physical Anthropology*, **44**, 113–22.

Perzigian, A. J. (1977). Fluctuating dental asymmetry: variation among skeletal populations. *American Journal of Physical Anthropology*, **47**, 81–8.

Perzigian, A. J. (1981). Allometric analysis of dental variation in a human population. *American Journal of Physical Anthropology*, **54**, 341–5.

Perizigian, A. J. (1984). Human odontometric variation: an evolutionary and taxonomic assessment. *Anthropologie*, **22**, 193–7.

Peters, D. S. (1983). Evolutionary theory and its consequences for the concept of adaptation. In *Dimensions of Darwinism*, ed. M. Grene, pp. 315–27. Cambridge: Cambridge University Press.

Pickford, M. (1986). Sexual dimorphism in Proconsul. *Human Evolution*, **1**, 111–48.

Pietrusewsky, M. (1980). Comment on Brace. *Current Anthropology*, **21**, 157.

Pilbeam, D. P. & Gould, S. J. (1974). Size and scaling in human evolution. *Science*, **186**, 892–901.

Pirie, P. L. (1978). Allometric scaling in the postcanine dentition with reference to primate diets. *Primates*, **19**, 583–91.

Portmann, A. (1967). *Animal Forms and Patterns*. New York: Schocken.

Post, D., Goldstein, S. & Melnick, D. (1978). An analysis of cercopithecoid odontometrics. II. Relations between dental dimorphism, body size dimorphism and diet. *American Journal of Physical Anthropology*, **49**, 533–44.

Potter, R. H. Y. (1972). Univariate versus multivariate differences in tooth size according to sex. *Journal of Dental Research*, **51**, 716–22.

Potter, R. H. Y., Alcazaren, A. B., Herbosa, F. M. & Tomaneng, J. (1981). Dimensional characteristics of the Filipino dentition. *American Journal of Physical Anthropology*, **55**, 33–42.

Potter, R. H. Y., Dahlberg, A. A., Merritt, A. D. & Connelly, P. M. (1968). Genetic studies of tooth size factors in Pima Indian families. *American Journal of Human Genetics*, **20**, 89–100.

Potter, R. H. Y. & Nance, W. E. (1976). A twin study of dental dimensions. I. Discordance, asymmetry and mirror imagery. *American Journal of Physical Anthropology*, **44**, 391–6.

Potter, R. H. Y., Nance, W. E., Yu, P. & Davis, W. B. (1976). A twin study of dental dimensions. II. Independent genetic determinants. *American Journal of Physical Anthropology*, **44**, 397–412.

Potter, R. H. Y., Rice, J. P., Dahlberg, A. A. & Dahlberg, T. (1983). Dental size traits within families: Path analysis for first molar and lateral incisors. *American Journal of Physical Anthropology*, **61**, 283–9.

Prakash, S., Kaul, V. & Kanta, S. (1979). Observations on the Bhutanese dentition. *Human Biology*, **51**, 23–30.

Preston, C. B. (1980). Comment on Brace. *Current Anthropology*, **21**, 157.

Provine, W. B. (1983). The development of Wright's theory of evolution: Systematics, adaptation and drift. In *Dimensions of Darwinism*, ed. M. Grene, pp. 43–70. Cambridge: Cambridge University Press.

Przibram, H. (1907). Vererbungsversuche über asymmetrische Angefarbung bei Angorakatzen. *Archiv Entwicklung*, **25**, 260–5.

Radnzic, D. (1987). Comparative study of mesiodistal diameters and arch diameters between indigenous British and Pakistani immigrant populations. *American Journal of Physical Anthropology*, **72**, 479–83.

Rao, C. R. (1952). *Advanced Statistical Methods in Biometric Research*. New York: Wiley.

Rao, D. C., Morton, N. E. & Yee, S. (1974). An analysis of family resemblance. II. Linear model for familial correlation. *American Journal of Human Genetics*, **26**, 331–59.

Rao, D. C., Morton, N. E. & Yee, S. (1976). Resolution of cultural and biological inheritance by path analysis. *American Journal of Human Genetics*, **28**, 228–48.

Rayner, J. M. W. (1985). Linear relations in biomechanics: the statistics of scaling functions. *Journal of Zoology*, **206**, 415–39.

Raup, C. M. (1968). Theoretical morphology of Ecinoid growth. *Journal of Paleontology*, **42**, 50–63.

Reeve, E. C. R. (1960). Some genetic tests on asymmetry of sternopleural chaeta number in Drosophila. *Genetic Research*, **1**, 151–72.

Releford, J. H. (1984). Morphological size and shape variation among local human populations. *Journal of Human Evolution*, **13**, 191–4.

Rensch, B. (1969). *Evolution Above the Species Level*. New York: Columbia University Press.

Reyment, R. A., Blackith, R. E. & Cambell, N. A. (1984). *Multivariate Morphometrics*. London: Academic Press.

Reynolds, R. T. (1972). Sexual dimorphism in accipiter hawks: a new hypothesis. *Condor*, **74**, 191–7.

Rice, J., Cloninger, C. R. & Reich, T. (1978). Multifactorial inheritance with cultural transmission. *American Journal of Human Genetics*, **30**, 618–43.

Rice, J., Cloninger, C. R. & Reich, T. (1980). General causal model for sex differences in familial transmission of multifactorial traits. *Social Biology*, **27**, 36–47.

Rice, J., Nicholls, P. L. & Gottesman, I. I. (1981). Assessment of sex differences for multifactorial traits using path analysis. *Psychology Research*, **4**, 301–12.

Richardson, E. R. & Malhotra, S. K. (1975). Mesiodistal crown dimensions of the permanent dentition of American Negroes. *American Journal of Orthodontics*, **68**, 157–64.

Riedl, R. (1978). *Order in Living Organisms*. New York: Wiley.

Riesenfeld, A. (1970). The effect of environmental factors on tooth development: an experimental investigation. *Acta Anatomica*, **77**, 188–215.

Riska, J. M. V. & Atchley, W. R. (1985). Genetics of growth prediction patterns of brain size evolution. *Science*, **229**, 668–70.

Robinson, J. T. (1954). Prehominid dentition and hominid evolution. *Evolution*, **8**, 326–34.

Robinson, J. T. (1956). The dentition of the *Australopithecinae*. *Memoirs of the Transvaal Museum*, **9**, 1–179.

Rose, C. (1906). Ueber die Rückbildung des seitlichen Scheiderzähne des Oberkiefers und Weisheitszähne in menschlichen Gebisse. *Monatschrift für Zahnheilkunde*, **24**, 225–58.

Rosenberg, E. (1895). Ueber Umformungen an den Incisiven der zweiten Zahngeneration des Menschen. *Morphologisches Jahrbuch*, **22**, 265–339.

Rosenzweig, K. A. & Zilberman, Y. (1967). Dental morphology of Jews from Yemen and Cochin. *American Journal of Physical Anthropology*, **26**, 15–22.

Rosenzweig, K. A. & Zilberman, Y. (1969). Dentition of Bedouin in Israel. *American Journal of Physical Anthropology*, **31**, 199–204.

Rutimeyer, F. (1863). Beitrage zur Kenntniss des fossilen Pferde und sur vergleicheuden Odontographie der Huftiere überkaupt. *Vehrnatür Gesellschaft*, **3**, 558–62.

Saayman, G. S. (1971a). Baboon's responses to predators. *African Wildlife*, **25**, 46–9.

Saayman, G. S. (1971b). Aggressive behaviour in free ranging Chacma baboons. *Journal of Behavioural Science*, **1**, 77–83.

Sakai, T., Hanamura, H. & Ohno, N. (1971). Tooth size of the Pashtun and Tajik in Afghanistan. *Journal of the Anthropological Society of Nippon*, **79**, 159–77.

Sakura, H. (1970). Dentition of the Amud man. In *The Amud Man and his Cave Site*, ed. H. Suzuki & F. Takai, pp. 207–30. Tokyo: University of Tokyo.

Sanin, C. & Savara, B. S. (1971). An analysis of permanent mesiodistal crown size. *American Journal of Orthodontics*, **59**, 488–500.

Schamschula, R. G., Barnes, D. E. & Adkins, B. L. (1972). Caries aetiology in Papua New Guinea. *Australian Dental Journal*, **17**, 188–95.

Schmalhausen, I. I. (1949). *Factors of Evolution*. Philadelphia: Blakiston.

Schmidt-Nielsen, K. (1975). Scaling in biology: the consequence of size. *Journal of Experimental Zoology*, **194**, 287–308.

Schmidt-Nielsen, K. (1984). *Scaling*. Cambridge: Cambridge University Press.

Schranze, D. (1964). Morphologische unterscheide zwischen männlicher und weiblicher Zähne. *Acta Morphologica*, **12**, 401–6.

Sciulli, P. W. (1979). Size and morphology of the permanent dentition in prehistoric Ohio Valley Amerindians. *American Journal of Physical Anthropology*, **50**, 615–28.

Sciulli, P. W., Doyle, W. J., Kelly, S., Siegel, P. & Siegel, M. I. (1979). The interaction of stressors in the induction of increased levels of fluctuating asymmetry in the laboratory rat. *American Journal of Physical Anthropology*, **50**, 279–84.

Scott, E. C. (1979). Increase in tooth size in prehistoric coastal Peru, 10 000–1000 BC. *American Journal of Physical Anthropology*, **50**, 251–8.

Scott, W. B. (1892). On variations and mutations. *American Journal of Science*, **48**, 355–74.

Seegmiller, R., Fraser, E. & Clarke, F. (1977). Mandibular growth retardation as a cause of cleft palate in mice homozygous for the chondrodysplasia gene. *Journal of Embryology and Experimental Morphology*, **38**, 227–38.

Seipel, C. (1946). Variation in tooth position: a metric study of variation and adaptation in the deciduous and permanent dentitions. *Swedish Dental Journal*, **39**, 1–176.

Selander, R. K. (1967). Sexual dimorphism and niche utilization in birds. *Condor*, **68**, 113–51.

Selander, R. K. (1972). Sexual selection and dimorphism. In *Sexual Selection and the Descent of Man*, 1871–1971, ed. B. Campbell, pp. 180–230. Chicago: Aldine.

Selmer-Olsen, R. (1949). An odontological study of the Norwegian Lapps. *Skrift Norske Vidensk-Akademi*, **3**, 1–167.

Selye, H. (1952). *The Story of the Adaptation Syndrome*. Montreal: Acta Medical Publishers.

Selye, H. (1973). The evolution of the stress concept. *American Scientist*, **61**, 692–9.

Shapiro, B. L. (1971). Developmental stability and instability. *Journal of Dental Research*, **50**, 1505–6.

Shapiro, B. L. (1975). Amplified developmental instability in Down's syndrome. *Annals of Human Genetics*, **38**, 429–37.

Sharma, J. C. (1983). Dental morphology and odontometry of Tibetan immigrants. *American Journal of Physical Anthropology*, **61**, 495–505.

Sharma, J. C. & Corruccini, R. S. (1987). Genetic variance of dental asymmetry in Punjabi twins. *American Journal of Physical Anthropology*, **72**, 253.

Shaw, J. C. M. (1931). *The Teeth, the Bony Palate and the Mandible in Bantu Races of South Africa*. London: J. Bale and Sons.

Shaw, J. H. & Griffiths, D. (1963). Dental abnormalities attributable to protein deficiency during reproduction. *Journal of Nutrition*, **80**, 123–41.

Shea, B. T. (1981). Relative growth of the limbs and trunk in the African apes. *American Journal of Physical Anthropology*, **56**, 179–202.

Shea, B. T. (1983a). Allometry and heterochromy in African apes. *American Journal of Physical Anthropology*, **62**, 275–89.

Shea, B. T. (1983b). Size and diet in the evolution of African ape craniodental form. *Folia Primatologica*, **40**, 32–68.

Shea, B. T. (1985a). Ontogenetic allometry and scaling: a discussion based on the growth and form of the skull in African apes. In *Size and Scaling in Primate Biology*, ed. W. L. Jungers, pp. 175–205. New York: Plenum.

Shea, B. T. (1985b). The ontogeny of sexual dimorphism in the African apes. *American Journal of Physical Anthropology*, **58**, 183–8.

Shea, B. T. (1986). Ontogenetic approaches to sexual dimorphism in Anthropoids. *Human Evolution*, **1**, 97–110.

Shea, B. T. & Gomez, A. M. (1988). Tooth scaling and evolutionary dwarfism: an investigation of allometry in human pygmies. *American Journal of Physical Anthropology*, **77**, 177–82.

Siegel, M. I. & Doyle, W. (1975a). The differential effects of prenatal and postnatal audiogenic stress on fluctuating dental asymmetry. *Journal of Experimental Zoology*, **191**, 211–14.

Siegel, M. I. & Doyle, W. (1975b). The effects of cold stress on fluctuating asymmetry in the dentition of the mouse. *Journal of Experimental Zoology*, **193**, 385–9.

Siegel, M. I., Doyle, W. & Kelly, C. (1977). Heat stress, fluctuating asymmetry and prenatal selection in the laboratory rat. *American Journal of Physical Anthropology*, **46**, 121–6.

Siegel, M. I. & Smookler, H. H. (1973). Fluctuating dental asymmetry and audiogenic stress. *Growth*, **37**, 35–9.

Simpson, G. G. (1953). *The Major Features of Evolution*. New York: Columbia University Press.

Smith, B. H., Garn, S. M. & Cole, P. E. (1982). Problems of sampling and inference in the study of fluctuating dental asymmetry. *American Journal of Physical Anthropology*, **58**, 281–9.

Smith, C. (1974). Concordance in twins: methods and interpretations. *American Journal of Human Genetics*, **26**, 454–64.

Smith, C. (1975). Quantitative inheritance. In *Textbook of Human Genetics*, ed. G Fraser & O. Mayo, pp. 382–441. Oxford: Blackwell.

Smith, J. M., Burian, R., Kauffman, P., Alberch, P., Campbell, J., Goodwin, B., Lande, R., Raup, D. & Wolpert, L. (1985). Developmental constraints and evolution. *Quarterly Review of Biology*, **60**, 265–87.

Smith, P. (1977). Selective pressures and dental evolution of hominids. *American Journal of Physical Anthropology*, **47**, 453–8.

Smith, P. (1982). Dental reduction: selection or drift? In *Teeth: Form, Function and Evolution*, ed. B. Kurten, pp. 366–81. New York: Columbia University Press.

Smith, P., Brown, T. & Wood, W. B. (1981). Tooth size and morphology in a recent Australian Aboriginal population from Broadbeach, S.E. Queensland. *American Journal of Physical Anthropology*, **55**, 423–32.

Smith, R. J. (1981a). Interspecific scaling of maxillary canine size and shape in female primates: relationships to social structure and diet. *Journal of Human Evolution*, **10**, 165–73.

Smith, R. J. (1981b). On the definition of variables in studies of primate dental allometry. *American Journal of Physical Anthropology*, **55**, 323–9.

Smith, R. J. (1985). The present as a key to the past: body weight of Miocene hominoids as a test of allometric methods for paleontological inference. In *Size and Scaling in Primate Biology*, ed. W. L. Jungers, pp. 437-48. New York: Plenum.

Sofaer, J. A. (1973). A model relating developmental interaction and differential evolutionary reduction of tooth size. *Evolution*, **27**, 427–34.

Sofaer, J. A. (1979). Human tooth-size asymmetry in cleft lip with or without cleft palate. *Archives of Oral Biology*, **24**, 141–6.

Sofaer, J. A., Bailit, H. L. & MacLean, C. J. (1971a). A developmental basis for differential tooth size reduction during hominid evolution. *Evolution*, **25**, 509–17.

Sofaer, J. A., Chung, C. S., Niswander, J. D. & Runck, D. W. (1971b). The developmental interaction of size and agenesis among maxillary incisors. *Human Biology*, **43**, 36–45.

Sokal, R. R. & Sneath, P. H. A. (1963). *Principles of Numerical Taxonomy*. San Francisco: Freeman.

Solow, B. (1966). The pattern of craniofacial associations. *Acta Odontologica Scandinavica*, **24**, 1–170.

Stahle, H. (1959). The determination of mesiodistal crown width of unerupted permanent cuspids and bicuspids. *Helvetia Odontologica Acta*, **3**, 13–17.

Stapleton, H. E. (1958). Ancient and modern aspects of Pythagoreanism. *Osiris*, **30**, 12–53.

Stein, H. (1981). *The Sick Black Child*. Johannesburg: Witwatersrand University Press.

Steyn, N. P., Albertse, E. C., Kotze, T. J. & van Heerden, L. (1986). Analysis of diets of 12 year old children. *South African Medical Journal*, **69**, 739–42.

Steudel, K. (1985). Allometric perspectives on fossil Catarrhine morphology. In *Size and Scaling in Primate Biology*, ed. W. L. Jungers, pp. 449–76. New York: Plenum.

Storer, R. W. (1966). Sexual dimorphism and food habits in three North American accipiters. *Auk*, **83**, 423–36.

Struhsaker, T. T. (1969). Correlates of ecology and social organization among African cercopitheicines. *Folia Primatologica*, **11**, 80–118.

Suarez, B. K. (1974). Neanderthal dental asymmetry and the probable mutation effect. *American Journal of Physical Anthropology*, **41**, 411–16.

Suarez, B. K. & Bernor, R. (1972). Growth fields in the dentition of the Gorilla. *Folia primatologica*, **18**, 356–9.

Suarez, B. K. & Williams, B. J. (1973). Dental growth fields and premolar morphology. *Journal of Dental Research*, **52**, 632–6.

Sumner, F. B. & Huestis, R. R. (1921). Bilateral asymmetry and its relation to certain problems of genetics. *Genetics*, **6**, 445–85.

Susman, R. L. & Creel, N. (1979). Functional and morphological affinities of the subadult hand from Olduvai Gorge. *American Journal of Physical Anthropology*, **51**, 311–32.

Suzuki, M. & Sakai, T. (1957). The living Sakhalin Ainu dentition. *Anthropological Reports, Nippon*, **18**, 303–46.

Takehisa, S. (1957). An anthropological and morphological comparative study of American White and Japanese teeth. *Shikagakuho*, **57**, 1–41.

Tanner, J. M., Prader, A., Habich, H. & Ferguson-Smith, M. A. (1959). Genes on the Y chromosome influencing the rate of maturation in man. *Lancet*, **2**, 141–4.

Taverne, P. P. (1980). Een Fysisch-Anthropologisch Onderzoek van de Gebitten van Vier Surinaamse Bevolkingsgroepen. Thesis, University of Groningen.

Thenius, E. (1970). Zur Evolution und Verbreitungsgeschichte der Suidae. *Zeitschrift für Saugetier*, **35**, 321–42.

Thoday, J. M. (1958). Homeostasis in a selection experiment. *Heredity*, **12**, 401–15.

Thompson, G. W., Anderson, D. L. & Popovich, F. (1975). Sexual dimorphism in dentition mineralization. *Growth*, **39**, 289–301.

Thompson, L. A. (1964). *A New Dictionary of Birds*. New York: McGraw-Hill.

Thorpe, R. S. & Leamy, L. (1984). Morphometric studies in inbred and hybrid house mice. *Journal of Zoology*, **199**, 421–32.

Tobias, P. V. (1955). Teeth, jaws and genes. *Journal of the Dental Association of South Africa*, **10**, 88–104.

Tobias, P. V. (1967). *Olduvai Gorge II: The Cranium and Maxillary Dentition of Australopithecus (Zinjanthropus) boisei*. Cambridge: Cambridge University Press.

Tobias, P. V. (1972). Progress and problems in the study of early man in sub-Saharan Africa. In *The Functional and Evolutionary Biology of Primates*, ed. R. H. Tuttle, pp. 63–93. Chicago: Aldine.

Tobias, P. V. (1975). Stature and secular trend among South African Negroes and San. *South African Journal of Medical Sciences*, **40**, 145–64.

Tobias, P. V. (1978). The secular trend in African peoples. In *Mankind and Medicine in the Third Millennium*, ed. A. J. Brink & C. M. Lewis, pp. 46–74. Stellenbosch: Stellenbosch University Press.

Tobias, P. V. (1985). The negative secular trend. *Journal of Human Evolution*, **14**, 347–56.

Tobias, P. V. & Netcher, D. (1976). Evidence from African Negro skeletons for a reversal of the usual secular trend. *Journal of Anatomy*, **121**, 435–6.

Tobias, P. V. & Netcher, D. (1977). Reversal of the usual secular trend, as evidenced by South African Negro crania and femora. *Mitteilungen Anthropologischen Gesellschaft, Wien*, **103**, 233–9.

Townsend, G. C. (1983a). Tooth size in children and young adults with trisomy 21. *Archives of Oral Biology*, **28**, 159–66.

Townsend, G. C. (1983b). Fluctuating dental asymmetry in Down's syndrome. *Australian Dental Journal*, **28**, 39–44.

Townsend, G. C. (1987). A correlative analysis of dental crown dimensions in individuals with Down syndrome. *Human Biology*, **59**, 537–48.

Townsend, G. C. & Alvesalo, L. (1985). Tooth size in 47 XYY males: evidence for a direct effect of the Y chromosome on growth. *Australian Dental Journal*, **30**, 268–72.

Townsend, G. C. & Brown, T. (1978a). Inheritance of tooth size in Australian Aborigines. *American Journal of Physical Anthropology*, **48**, 305–14.

Townsend, G. C. & Brown, T. (1978b). Heritabilities of permanent tooth size. *American Journal of Physical Anthropology*, **49**, 497–502.

Townsend, G. C. & Brown, T. (1979a). Tooth size characteristics of Australian Aborigines. In *Occasional Papers in Human Biology*, pp. 17–38. Canberra: Australian Institute of Aboriginal Studies.

Townsend, G. C. & Brown, T. (1979b). Family studies of tooth size factors in the permanent dentition. *American Journal of Physical Anthropology*, **50**, 183–90.

Townsend, G. C. & Brown, T. (1980). Dental asymmetry in Australian Aboriginals. *Human Biology*, **52**, 661–73.

Townsend, G. C. & Garcia-Godoy, F. (1984). Fluctuating asymmetry in the deciduous dentition of Dominican mulatto children. *Archives of Oral Biology*, **29**, 483–6.

Trinkaus, E. (1983). *The Shanidar Neanderthals*. New York: Academic Press.

Turner, C. G. & Swindler, D. R. (1978). The dentition of the New Britain West Nakanai Melanesians. *American Journal of Physical Anthropology*, **49**, 361–72.

Unterhalter, B. (1982). Inequalities of health and disease, the case of mortality rates for the city of Johannesburg. *International Journal of Health Services*, **12**, 617–36.

Utermohle, C. J. (1983). Cranial Variation of the Eastern Eskimo. Ph.D. Thesis, Arizona State University, Tempe: Arizona.

Utermohle, C. J., Zegura, S. L. & Heathcote, G. M. (1983). Multiple observers, humidity and the choice of precision statistics. *American Journal of Physical Anthropology*, **61**, 85–95.

van Reenen, J. F. (1961). The use of the concept of tooth material as an indicator of tooth size in a group of Kalahari Bushmen. *South African Journal of Science*, **57**, 347–52.

van Reenen, J. F. (1964). Dentition, jaws and palate of the Kalahari Bushman. *Journal of the Dental Association of South Africa*, **19**, 38–44.

van Reenen, J. F. (1966). Dental features of a low-caries primitive population. *Journal of Dental Research*, **45**, 703–13.

van Reenen, J. F. (1982). The effects of attrition on tooth dimensions of San (Bushmen). In *Teeth: Form, Function and Evolution*, ed. B. Kurten, pp. 182–203. New York: Columbia University Press.

van Valen, L. (1961). Growth fields in the dentition of Peromyscus. *Evolution*, **15**, 272–7.

van Valen, L. (1962). A study of fluctuating asymmetry. *Evolution*, **16**, 125–42.

van Valen, L. (1963). Selection in a natural population of *Merychippus primus*. *Nature*, **197**, 1181–3.

van Valen, L. (1970). An analysis of developmental fields. *Developmental Biology*, **23**, 456–77.

186    *References*

Waddington, C. H. (1940). *Organisers and Genes*. Cambridge: Cambridge University Press.
Waddington, C. H. (1942). Canalization of development and the inheritance of acquired characters. *Nature*, **150**, 563–5.
Waddington, C. H. (1957). *The Strategy of the Genes*. New York: Macmillan.
Walimbe, S. R. (1985). Crown dimensions among muslim males of Ramapuram, Andhra Pradesh. In: *Dental Anthropology: Application and Methods*. Ed. V. R. Reddy, pp. 331–2. New Delhi: Inter-India Publications.
Walker, A. R. P. (1966). Nutritional, biochemical and other studies on South African populations. *South African Medical Journal*, **44**, 814–52.
Walker, A. R. P. (1982). Breakfast habits of adolescents in four South African populations. *American Journal of Clinical Nutrition*, **36**, 450–656.
Wallace, J. (1978). Evolutionary trends in early hominid evolution. In *Early Hominids in Africa*, ed. C. J. Jolly, pp. 285–310. London: Duckworth.
Wessels, N. K. (1982). A catalogue of processes responsible for Metazoan morphogenesis. In *Evolution and Development*, ed. J. T. Bonner, pp. 115–54. Berlin: Springer.
Wiedersheim, R. (1895). *The Structure of Man*. London: Methuen.
Wilder, H. H. (1904). Duplicate twins and duplicate monsters. *American Journal of Anatomy*, **3**, 387–472.
Williams, R. C. (1978). The probable mutation effect: Neutral alleles and structural reduction. *Human Biology*, **50**, 173–81.
Wolpoff, M. H. (1969). The effect of mutations under conditions of reduced selection. *Social Biology*, **16**, 11–23.
Wolpoff, M. H. (1971). *Metric Trends in Hominid Dental Evolution*. Cleveland: Case Western Reserve University Press.
Wolpoff, M. H. (1978a). Some aspects of canine size in the Australopithecines. *Journal of Human Evolution*, **7**, 115–26.
Wolpoff, M. H. (1978b). Analogies and interpretation in palaeoanthropology. In *Early Hominids in Africa*, ed. C. J. Jolly, pp. 641–704. London: Duckworth.
Wolpoff, M. H. (1985). Tooth size – body size scaling in a human population: Theory and practice of an allometric analysis. In *Size and Scaling in Primate Biology*, ed. W. L. Jungers, pp. 273–318. New York: Plenum.
Wood, B. A. (1979). Models for assessing relative canine size in fossil hominids. *Journal of Human Evolution*, **8**, 493–502.
Wood, B. A. & Stack, C. G. (1980). Does allometry explain the differences between gracile and robust Australopithecines? *American Journal of Physical Anthropology*, **52**, 55–62.
Woolf, C. M. & Gianas, A. D. (1976). Congenital cleft lip and fluctuating dermatoglyphic asymmetry. *American Journal of Human Genetics*, **28**, 400–3.
Woolf, C. M. & Gianas, A. D. (1977). A study of fluctuating dermatoglyphic asymmetry in the sibs and parents of cleft lip propositi. *American Journal of Human Genetics*, **29**, 503–7.
Wortman, J. L. (1886). *Comparative Anatomy of the Vertebrates*. New York: Annals of Systematic Dentistry.
Wright, S. (1931). Evolution in Mendelian populations. *Genetics*, **16**, 97–159.
Wright, S. (1932). The roles of mutation, inbreeding, crossbreeding and selection. *Proceedings of the 6th International Congress of Genetics*, **1**, 356–66.
Wright, S. (1967). Comments on the preliminary working papers of Eden and Waddington. In *Mathematical Challenges to the Neo-Darwinian Interpretation of*

*Evolution*, ed. P. S. Moorhead & M. M. Kaplan, pp. 117–20. Philadelphia: Wistar Press.

Wright, S. (1980). Genetic and organismic selection. *Evolution*, **34**, 825–43.

Wright, S. (1982). The shifting balance theory and macroevolution. *Annual Review of Genetics*, **16**, 1–19.

Xiang-Qing, S. (1981). Comment on Brace. *Current Anthropology*, **22**, 560.

Yamada, H., Kogiso, T. & Liao, J. Y. (1986). Correlation matrices for mesiodistal and buccolingual crown diameters in Japanese and Chinese permanent teeth. *Journal of the Anthropological Society of Nippon*, **94**, 473–9.

Y'Edynak, G. (1978). Culture, diet and dental reduction in Mesolithic forager fishers of Yugoslavia. *Current Anthropology*, **19**, 616–18

Yerkes, R. M. (1901). A study of variation in the fiddler crab *Gelasimus pugulator*. *Proceedings of the American Academy of Arts and Sciences*, **36**, 415–42.

Zelditch, M. L. (1988). Ontogenetic variation in patterns of phenotypic integration in the laboratory rat. *Evolution*, **42**, 28–41.

Zihlman, A. L. & Cramer, D. L. (1978). Skeletal differences between pygmy (*Pan paniscus*) and common (*Pan troglodytes*) chimpanzees. *Folia Primatologica*, **29**, 86–94.

# Index